SCIENCE
A CLOSER LOOK
Reading Essentials

Macmillan/McGraw-Hill

Macmillan/McGraw-Hill

Contents

CHAPTER 1

Kingdoms of Life

Vocabulary

cell the smallest unit of living matter

organism a living thing that carries out five basic life functions

tissue a group of cells that do the same job

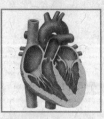

organ a group of tissues that work together to do a job

organ system a group of organs that work together to do a job

trait a feature of a living thing

kingdom the largest group into which a living thing can be classified

root the part of a plant that takes up water and minerals from the ground

stem the part of a plant that holds the plant upright

photosynthesis a process in plants that uses energy from sunlight to make food

The Big Idea

What are living things and how are they classified?

 spore a cell in a seedless plant that can grow into a new plant

 **seed** a plant that is not fully grown

 **reproduction** the making of new living things

 **ovary** a part of a plant that stores egg cells

 **pollination** the process of moving pollen

 **fertilization** the male sex cell joining with the female sex cell to form a seed

 **germination** when a seed sprouts into a new plant

 **life cycle** the stages of growth and change that a living thing goes through

What are living things?

Plants and animals are living things. Plants and animals have cells (SELZ). A **cell** is the smallest unit of life. You are made of cells!

Living Things Have Needs

All living things have needs. They need water, food, and a place to live. Most also need oxygen (OK•suh•juhn)—a gas in air and water.

Living Things Reproduce

Living things are called organisms. An **organism** is a living thing that carries out life functions. One function, or job, is to make more of one's own kind. Look at the birds below. The chick is their baby. To make more of one's own kind is to reproduce.

▲ Living things reproduce.

Other Life Functions

All organisms grow. Living things need energy to grow. They get energy from food. Then they must get rid of waste.

Organisms also respond to changes in the world around them. The sunflowers in the photo are all growing toward light.

▲ Living things respond to changes.

Is It a Living Thing?			
Life Function	Lizard	Rock	Car
Does it grow?	✓	✗	✗
Does it use fuel to get energy?	✓	✗	✓
Does it get rid of wastes?	✓	✗	✓
Does it reproduce?	✓	✗	✗
Does it react to changes in its environment?	✓	✗	✗

Read a Table

What are the three functions a car cannot do?

_____

_____

_____

✓ Quick Check

1. What do living things need?

_____.

_____.

How do plant and animal cells compare?

Cells have parts that help them live. Plants and animals have some cell parts that are the same. They have some parts that are different.

Most plant cells have chloroplasts (KLAWR•uh•plasts). These are filled with a green substance called chlorophyll (KLAWR•uh•fil). This substance helps the plant use the Sun's energy to make food.

Plant cells have stiff cell walls. Cell walls give the cell a boxlike shape. Animal cells do not have cell walls.

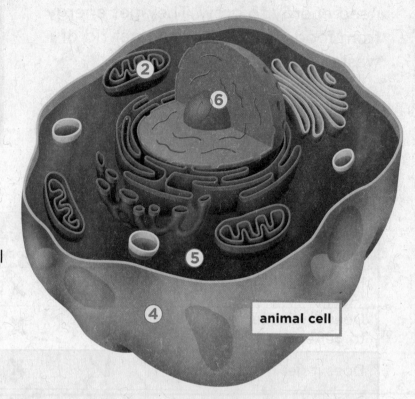

animal cell

① **cell wall**
This stiff structure protects and supports the plant cell.

② **mitochondrion** (my•tuh•KON•dree•uhn)
Food is burned here to give the cell energy.

③ **chloroplast**
The plant cell's food factory has chlorophyll.

④ **cell membrane**
This thin covering surrounds the cell. In plants, it is inside the cell wall.

⑤ **cytoplasm** (SY•tuh•plaz•uhm)
Filling the cell is a substance that is like jelly. It is mostly water. It also has important chemicals.

⑥ **nucleus** (NEW•klee•uhs)
This controls all cell activities.

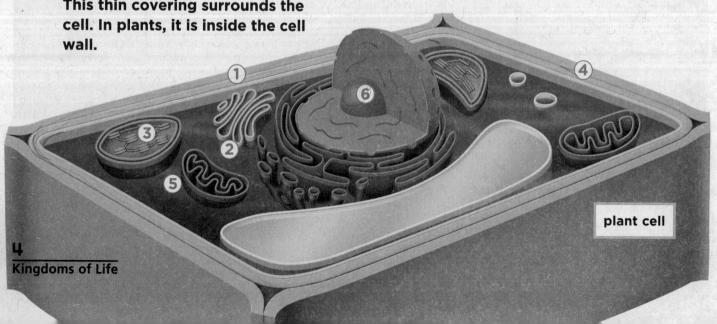

plant cell

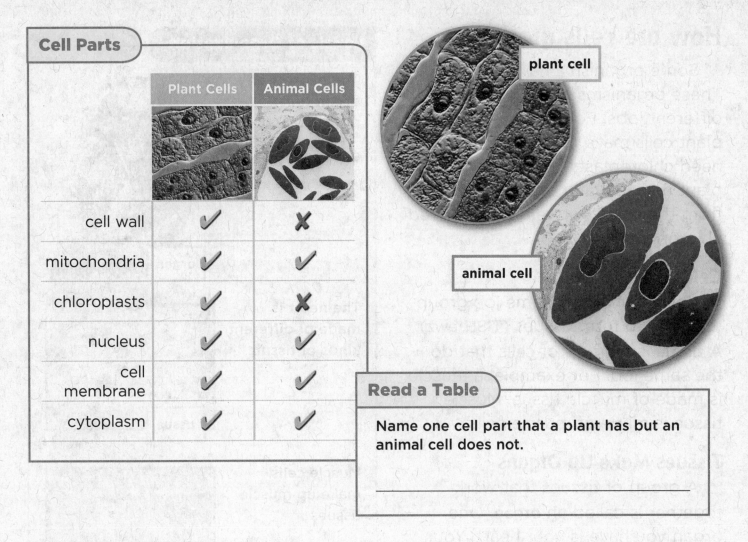

Cell Parts

	Plant Cells	Animal Cells
cell wall	✔	✘
mitochondria	✔	✔
chloroplasts	✔	✘
nucleus	✔	✔
cell membrane	✔	✔
cytoplasm	✔	✔

plant cell

animal cell

Read a Table

Name one cell part that a plant has but an animal cell does not.

_____

✅ **Quick Check**

Fill in the chart with the names of the correct cell parts.

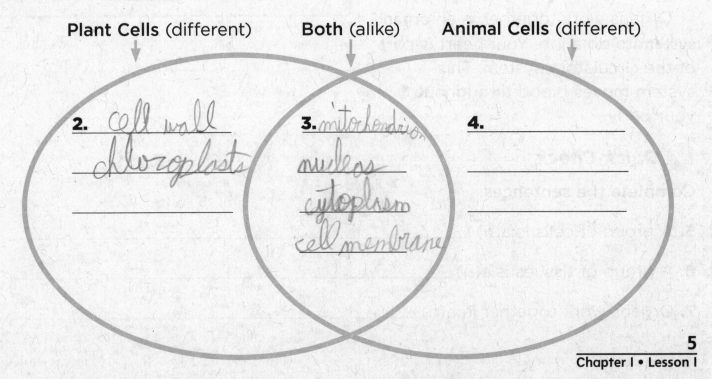

Plant Cells (different)   **Both** (alike)   **Animal Cells** (different)

2. Cell wall
   chloroplasts

3. mitochondrion
   nucleas
   cytoplasm
   cell membrane

4. _____

How are cells grouped?

Some organisms have many cells. These organisms have cells that do different jobs. For example, some plant cells make food. These cells need chloroplasts. Root cells in the plant take in water and nutrients from the soil. Root cells do not need chloroplasts.

Cells Make Tissues

Cells that do the same job group together to form tissues (TISH•ewz). A **tissue** is a group of cells that do the same job. For example, a muscle is made of muscle tissue. Muscle tissue is made of muscle cells.

Tissues Make Up Organs

A group of tissues that work together is called an **organ**. One organ you have is your heart. Your heart's job is to pump blood.

Organs Make Up Organ Systems

Organs work together in an **organ system** to do a job. Your heart is part of the circulatory system. This system moves blood throughout your body.

Levels of Organization

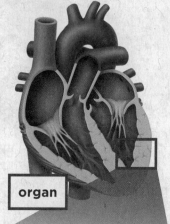

The heart is an organ that pumps blood.

organ

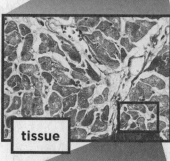

The heart is made of different kinds of tissue.

tissue

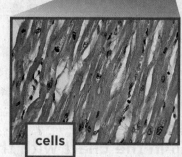

Muscle cells make up muscle tissues.

cells

✔ Quick Check

Complete the sentences.

5. A group of cells is a(n) _____.

6. A group of tissues is a(n) _____.

7. Organs work together in a(n) _____.

How can you see cells?

Most cells are too small to see with your eyes alone. Bacteria (bak•TEER•ee•uh) are the smallest cells of all.

Microscopes

You need a microscope (MIGH•kruh•skohp) to see most cells. A microscope is a tool that makes small things look much bigger. Scientists use microscopes to look at cells such as bacteria.

Microscopes are used to look at viruses, too. Viruses are even smaller than bacteria. Viruses cannot reproduce on their own. They need living cells to make more viruses.

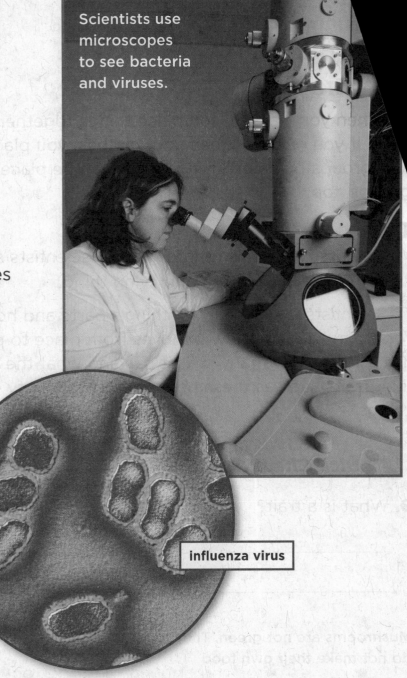

Scientists use microscopes to see bacteria and viruses.

influenza virus

✔ Quick Check

8. What does a microscope do?

_____

_____

 **e-Review** Summaries and quizzes online at www.macmillanmh.com

ng Things

ving things classified?

...u classify, you place things together that are
...ou need to clean your clothes, you place them
...ups by color. Living things can be placed into
...s, too.

...aits

To classify organisms into groups, scientists study
traits. A **trait** is a feature of a living thing.

Scientists look at a living thing's parts and how it gets
food. They look at how it moves from place to place.
They study how many cells it has and what the cells are
like. Traits such as these help scientists classify living
things.

✔ Quick Check

9. What is a trait?

_____

_____

Mushrooms are not green. They
do not make their own food. ▼

Classifying Organisms

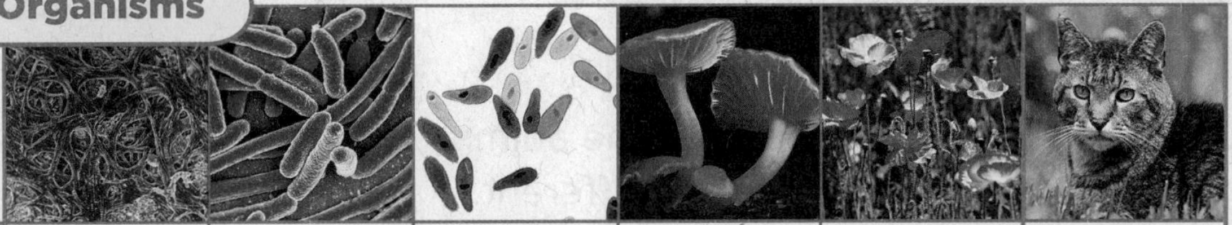

Kingdom	ancient bacteria	bacteria	protists	fungi	plants	animals
Number of cells	one	one	one or many	one or many	many	many
Nucleus	no	no	yes	yes	yes	yes
Food	make their own or get food from other organisms	make their own or get food from other organisms	make their own or get food from other organisms	get food from other organisms	make their own food	get food from other organisms
Move from place to place	yes	yes	yes	no	no	yes

Read a Chart

Which kingdoms have members that can move from place to place?

____

____

____

____

Six Kingdoms

Scientists classify living things into six kingdoms. A **kingdom** is the largest group into which living things can be classified. All members of a kingdom have the same basic traits.

Plants have their own kingdom. Animals do, too. Bacteria have two kingdoms. The chart above describes all six kingdoms.

Quick Check

10. Name the six kingdoms of living things.

_________________________, _________________________,

_________________________, _________________________,

_________________________, _________________________

How are organisms grouped within a kingdom?

Squirrels and lizards belong to the animal kingdom. However, they are very different. The animal kingdom can be divided into smaller groups.

A phylum (FIGH•luhm) is a smaller group within a kingdom. A phylum is broken down into smaller groups called classes. Classes have smaller groups called orders. Orders have families.

The chart shows these groups from largest to smallest. The smaller the group, the more the members have in common. The smallest groups are genus (JEE•nuhs) and species (SPEE•sheez).

The Eastern red squirrel is a member of the animal kingdom. ▼

Kingdom

Members of the animal kingdom move, eat food, and reproduce.

Phylum

Members of this phylum share at least one major characteristic, such as having a backbone.

Class

Members of this class produce milk for their young.

Order

Members of this order have long and sharp front teeth.

Family

Members of this family have a bushy tail.

Genus

Members of this genus climb trees.

Species

A species is made up of only one type of organism.

backbone

produce milk

long, sharp front teeth

bushy tail

climb trees

brown back, white front

▲ Lizards share the same phylum as squirrels. They both have backbones.

✔ Quick Check

11. Circle the grouping with organisms that have the most in common.

phylum   order   genus

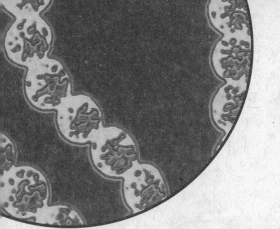

▲ This kind of bacteria causes strep throat.

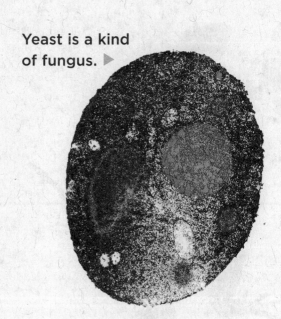

Yeast is a kind of fungus. ▶

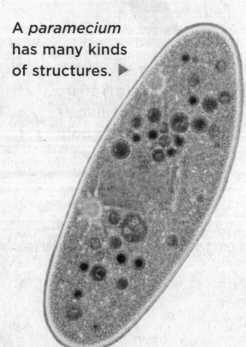

A *paramecium* has many kinds of structures. ▶

What kinds of organisms have only one cell?

The tiny organisms pictured on this page are microorganisms (migh•kroh•AWR•guh•niz•uhmz). Microorganisms are too small to be seen with just our eyes. Most are made of only one cell. You can find them in lakes, oceans, ponds, and rivers. You can also find them on your food, your skin, and even the pages of this book!

Bacteria

Bacteria are the smallest microorganisms. They are also the simplest. They have no cell nucleus. Some bacteria make their own food. Some bacteria get food from dead plants and animals.

Fungi

Fungi are living things that have some traits of plants. For example, their cells have a cell wall. However, unlike plants, fungi cannot make their own food.

Protists

Members of the protist kingdom have a cell nucleus. They also have different cell parts that do different jobs.

Some protists can cause disease, but most are harmless.

✅ Quick Check

12. How are fungi cells like plant cells?

_____

13. What are the smallest microorganisms?

_____

How are organisms named?

Scientists have rules for naming organisms. Each kind of organism has its own name. Each name shows how the organism is classified. The first part of the name is its genus. The second part is its species. This helps scientists identify living things. It also helps them study living things.

Genus and Species

Wolves, dogs, and coyotes all belong in the genus *Canis* (KAY•nuhs). Animals in *Canis* look alike. They all eat meat. But the species in this genus have different traits. One trait is color. Red wolves are *Canis rufus.* Gray wolves are *Canis lupus.*

✔ **Quick Check**

14. What does an animal's name show?

_____

_____

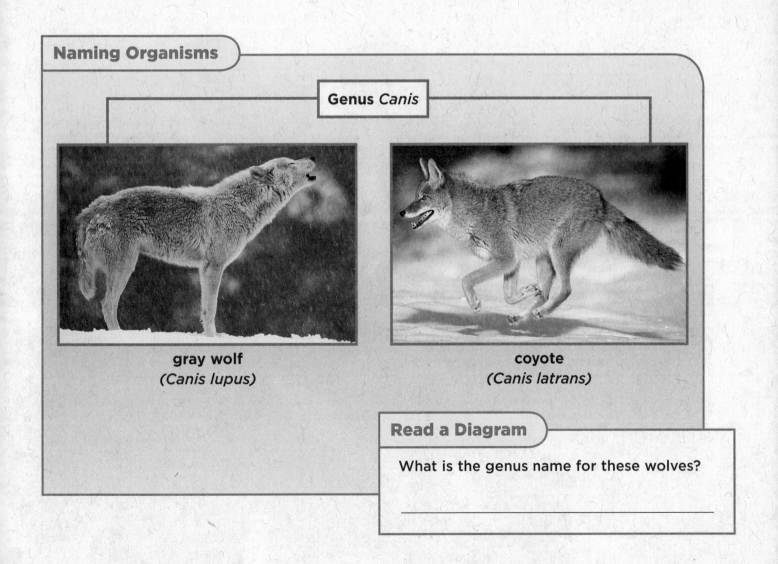

Naming Organisms

Genus *Canis*

gray wolf
(*Canis lupus*)

coyote
(*Canis latrans*)

Read a Diagram

What is the genus name for these wolves?

_____

How do we classify plants?

Plants come in many sizes, shapes, and colors. In all, there are about 400,000 different kinds of plants.

Leaves, Stems, and Roots

One way to classify plants is by their parts. Scientists look at the leaves, stems, and roots. Some plants do not have these parts. Plants can be classified into two groups. One group has leaves, stems, and roots. The other does not.

Wort plants do not have leaves, stems, or roots. ▶

The viola plant has roots, stems, and leaves. Its offspring grow from seeds. ▶

Seeds or Spores

Plants can make new plants in different ways. Some plants grow from seeds. Some plants grow from spores. Plants can be classified by how they make new plants.

Most plants you know about grow from seeds. Most plants that have roots, stems, and leaves also have seeds and fruit.

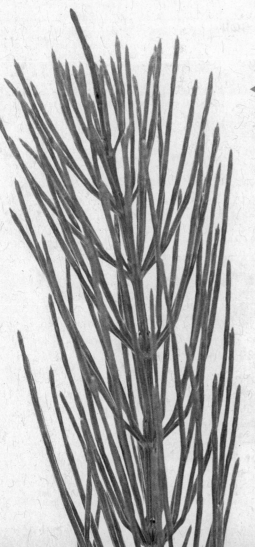

◀ The horsetail plant has roots, stems, and leaves, but no seeds.

✓ Quick Check

15. What are two ways we classify plants?

_____

_____

How do plants get what they need?

A scientist named Jan van Helmont wanted to know how plants meet their needs. He planted a small plant in a pot of soil. After five years, the small plant became a small tree. Only a small amount of soil was missing from the pot. Where did the plant get what it needed to grow?

We now know that plants make food from carbon dioxide, a gas in air. They also need water and energy from sunlight. Plants get what they need from carbon dioxide, water, and sunlight. These are the key ingredients that they need in order to grow.

The Role of Roots

Roots take up water and nutrients from the ground. They also hold plants in the soil. Some roots store food.

Different plants have different kinds of roots. Carrots have one large root called a taproot. Grasses have fibrous (FIGH•bruhs) roots that spread out into the soil.

tap root   fibrous root

The Role of Stems

A plant's stem grows above the ground. The **stem** moves food, water, and nutrients throughout the plant.

There are two kinds of stems. Most trees and shrubs have woody stems. Woody stems protect the plant. They also give it extra support. Smaller plants have stems that are soft, green, and can bend.

This woody stem is strong. It protects the plant. ▶

This soft, green stem can bend. ▶

✔ Quick Check

Match the following.

16. _____ take water and nutrients from soil.   **a.** Stems

17. _____ move food throughout the plant.   **b.** Roots

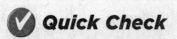

Why are leaves important?

Like all living things, plants need energy. They get their energy from the food they make.

Photosynthesis

Photosynthesis (foh•toh•SIN•thuh•suhs) is the process plants use to make food. It begins when sunlight hits the leaf. Light energy goes into the leaf cells. It enters the chloroplasts. There, the chlorophyll collects the light energy.

When the chloroplasts have enough energy, a change takes place. The plant uses the energy to combine water and carbon dioxide to make sugars, or food. Oxygen leaves the cells as waste.

Photosynthesis

Leaves take in sunlight.

Plants give off oxygen.

Leaves take in carbon dioxide from the air.

Roots take in water and nutrients from the soil.

Read a Diagram

What two things do leaves take in during photosynthesis?

_____

_____

LOG ON *Science in Motion* Watch how photosynthesis occurs at www.macmillanmh.com

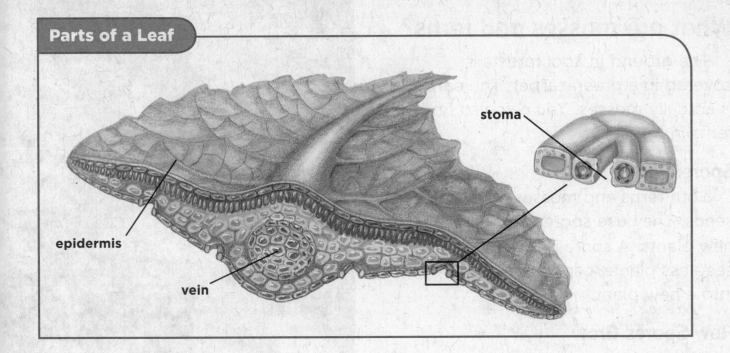

Parts of a Leaf

stoma

epidermis

vein

Collecting Carbon Dioxide

Carbon dioxide enters the leaves through tiny holes on the bottom of each leaf. The holes are called stomata (stoh•MAH•tuh). One hole is called a stoma (STOH•muh).

Collecting Water

The roots of a plant take up water from the ground. Small tubes called veins (VAYNZ) carry the water from the roots to the stems. They also move the water up the stem into each leaf.

✅ Quick Check

18. How does carbon dioxide enter a leaf?

_____

19. What carries water through a plant?

_____

What are mosses and ferns?

The ground in cool forests is covered in a green carpet. The carpet is actually mosses. You can also find ferns in forests.

Spores

Both ferns and mosses have no seeds. They use spores to make new plants. A **spore** is a cell in a seedless plant. A spore can grow into a new plant.

How Spores Grow

Spores grow inside spore cases. The cases protect the spores from heat. Spore cases also keep a plant from drying out. When the cases open, the spores are set free. Spores need light, nutrients, and water to grow into a new plant.

✓ **Quick Check**

20. What do spores need to grow?

_____

_____

▲ Mosses use spores to make new plants.

You can see the spore cases on the underside of fern leaves. ▶

How do we use plants?

Plants give us the food we eat. They are useful in many other ways, too.

Fruits and Vegetables

Fruits and vegetables are plants. Vegetables do not have seeds. Fruits have seeds. Not all fruit is sweet. Tomatoes and cucumbers are fruits.

Medicines and More

Some medicines come from plants. We use trees to build houses and furniture. We also use plants to make clothes. We even burn plants for fuel.

 Quick Check

21. What are two uses for plants, besides food?

_____

_____

Plants People Eat

Read a Photo

Complete the chart. Give an example of each.

Fruit (plant with seeds)	Vegetable (plant without seeds)
_____	_____

LOG ON e-Review Summaries and quizzes online at www.macmillanmh.com

How do we classify seed plants?

Grasses, trees, and some other plants grow from seeds. A **seed** contains a plant that is not fully grown. It also contains food for the plant. The coating of a seed keeps the plant inside safe.

Comparing Seeds and Parts

Seeds come in many shapes and sizes. We can use these differences to classify seed plants. A seed might be flat and slippery. It might be hard and round. A seed can also be small and smooth.

Plants that make seeds also have roots, stems, and leaves. We can also classify seed plants by their parts. Coconuts are giant seeds that come from palm trees. Palm trees have tall, woody stems. Watermelons and grapes grow on vines.

Each seed in a watermelon can make a new plant. ▼

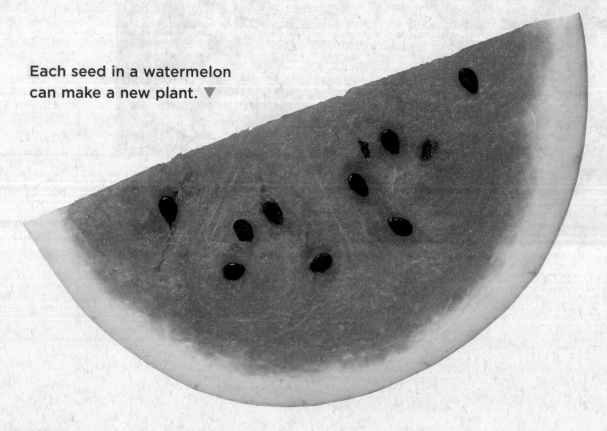

Flowers and Cones

We can also classify seed plants by where they store seeds. Most seed plants have flowers that make fruit. The fruit protects the seed inside.

Have you ever seen a pinecone? Pinecones come from conifers (KON•uh•furz). A conifer is a seed plant that has no flowers or fruit. Pine trees and other conifers make seeds on cones. There are male and female cones. The male cone is smaller.

The male cone makes a yellow powder called pollen. Wind can blow the pollen from a male cone to a female cone. A seed forms when pollen lands on the female cone.

Conifer Seeds

male

female

Read a Photo

Which pinecone makes pollen?

_____

◀ Coconuts are giant seeds from palm trees.

✓ Quick Check

22. A plant that is not fully formed is a

_____.

23. The yellow powder that is made by male cones is called _____.

How do seeds form?

Flowers are pretty, but they do not make food for a plant. They have a special job.

Reproduction

All living things can make more of their own kind. **Reproduction** (ree•pruh•DUK•shuhn) is how living things make new living things.

Some plants use flowers to reproduce. Flowers have male and female parts. The male part is the stamen. Inside the stamen is the anther. The anther makes pollen. It has male sex cells. The female part is the pistil. The pistil makes female sex cells, or eggs. The **ovary** inside the pistil stores the eggs.

The bright colors and sweet smells of flowers attract bees. ▼

Pollination

Wind can blow pollen from an anther to a pistil. Some plants rely on birds, insects, or other animals to move pollen.

For example, bees feed on nectar inside a flower. Nectar is a sugary liquid. If the bee touches an anther, pollen sticks to the bee's body and legs. At the next flower, the pollen falls off the bee onto that flower's pistil. This is called **pollination**.

Fertilization

Next, the pollen travels down a tube into the ovary. In the ovary, the male sex cell combines with an egg. Fertilization occurs when this happens. **Fertilization** is the process that forms a seed.

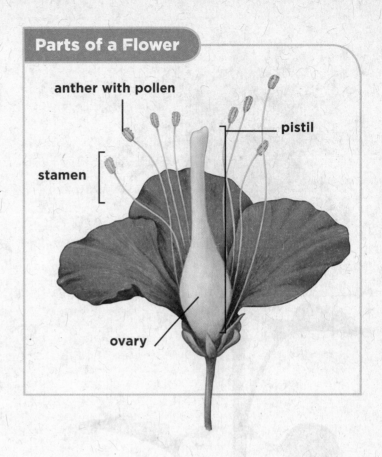

Parts of a Flower

anther with pollen

pistil

stamen

ovary

 Quick Check

24. Describe one way plants depend on animals.

_____

How do seeds grow?

The inside of a seed has a tiny plant. It also has food for the plant.

Germination

When conditions are right, the seed will grow. A root pushes through a crack in the seed. Then, a tiny stem grows upward. One or two leaves appear on the stem. This process is called **germination** (juhr•muh•NAY•shuhn).

The young plant is now a seedling. Seedlings need water, light, and nutrients to grow. If its needs are met, the seedling will grow into an adult plant. Then it can make its own seeds.

Seed Germination

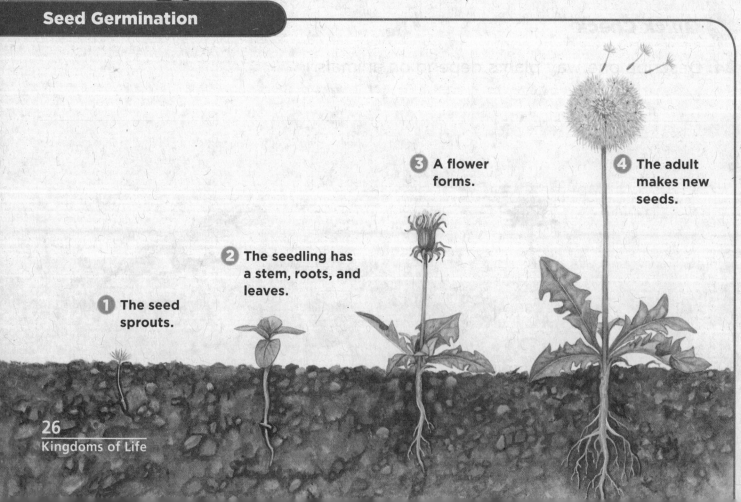

3 A flower forms.

4 The adult makes new seeds.

2 The seedling has a stem, roots, and leaves.

1 The seed sprouts.

Life Cycle of a Seed Plant

germination

adult plant

pollination

animal eats fruit

fruit with seeds

fertilization

Life Cycles

Some plants grow into adults in just days. For others, it can take years. The diagram shows the life cycle of a berry plant. A **life cycle** is how a living thing grows and reproduces.

✓ Quick Check

25. In a seed plant, what must happen before fertilization can take place?

_____

_____

How are plants alike and different from their parents?

Seeds grow into plants that look like their parents. However, they are not exact copies. The foxglove plant may have tall stems just like its parents. But the color of the flowers may be different.

Inherited Traits

Color, size, and shape are inherited (in•HER•uh•tid) traits. Inherited traits are traits passed from parent to offspring. When traits from two parents combine, the new plant may look different than its parents.

The flowers of a foxglove plant may have different colors. ▼

✅ Quick Check

26. Give an example of a trait that is inherited from a parent.

Choosing Traits

Farmers can use inherited traits to make better plants. If you want to grow a giant pumpkin, you would start with plants that made large pumpkins. Then you could move pollen from one pumpkin plant to the other.

When you do this, the pumpkin's offspring may be bigger. Many farmers pollinate their plants. They grow plants that many people like. Seedless watermelons are an example.

Making Bigger Pumpkins

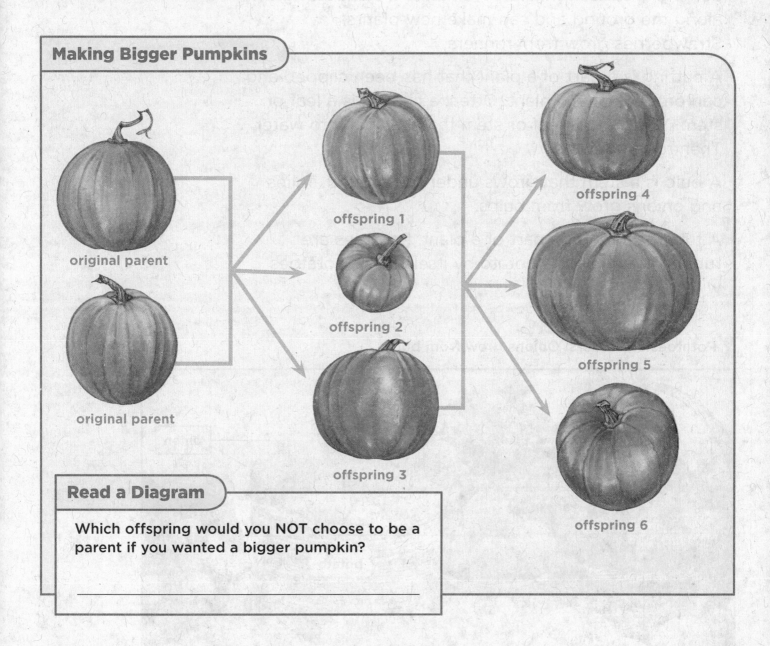

original parent

original parent

offspring 1

offspring 2

offspring 3

offspring 4

offspring 5

offspring 6

Read a Diagram

Which offspring would you NOT choose to be a parent if you wanted a bigger pumpkin?

_____

 **Quick Check**

27. Farmers _____ plants to make food that many people like.

What are other ways plants can reproduce?

Runners, Cuttings, Bulbs, and Tubers

Not all plants reproduce using flowers, cones, or spores. Plants can reproduce in other ways, too.

- Some plants use runners. A runner is a stem that grows along the ground and can make new plants. Strawberries grow from runners.

- A cutting is a part of a plant that has been clipped and can produce a new plant. Often, a cutting is a leaf or stem. You put the leaf or stem that was cut into water. Then roots grow.

- A bulb is a stem that grows under the ground. Tulips and onions grow from bulbs.

- A tuber is a storage part of a plant. Potatoes are tubers. If you plant a potato by itself, more potatoes will grow.

▼ Potatoes are tubers. Onions grow from bulbs.

✓ Quick Check

Complete the sentences.

28. A stem that grows under the

ground is a _____.

29. A part of a plant that has been

clipped is a _____.

30. A stem that grows along the

ground is a _____.

31. A storage part of a plant is a

_____.

Reproduction Without Seeds

daffodil

chain plant

Read a Photo

Match the plant to how the plant reproduces.

____ daffodil   **a.** bulb

____ chain plant   **b.** cutting

Kingdoms of Life

Match the vocabulary words to their definitions. Then find and circle those words in the puzzle at the bottom.

1. _____ a living thing

2. _____ a plant that has not started growing

3. _____ the smallest unit of living matter

4. _____ a feature of a living thing

5. _____ the process of moving pollen

6. _____ the largest group into which a living thing can be classified

7. _____ a part of a plant that stores egg cells

8. _____ the process that plants use to make food

9. _____ the male sex cell joins with the female sex cell

10. _____ the making of new living things

a. cell
b. organism
c. pollination
d. seed
e. fertilization
f. ovary
g. reproduction
h. trait
i. kingdom
j. photosynthesis

```
O R G A N I S M K S A R Q T Q
Q N H G C E L L C I E M Q R N
O V A R Y Q C G L U N E E A U
P O L L I N A T I O N G D I X
M F U D N A R O H U F R D T U
Y B L R T N V V   Q I B G O O
H R E P R O D U C T I O N Y M
F E R T I L I Z A T I O N J E
A G T O O R J V H C A K E Y K
P H O T O S Y N T H E S I S T
```

Use a word from the box to name each example described below.

organism

tissue

organ

spore

germination

life cycle

1. the stages of growth and change that a living

thing goes through _____

2. a cell in a seedless plant that can grow into a

new plant _____

3. a group of cells that do the same job _____

4. a living thing that carries out five basic life functions

on its own _____

5. when something begins to grow, as when a seed

sprouts into a new plant _____

6. a group of tissues that work together to do a job _____

Answer the question. Use at least one word from the box at the top of the page.

7. Name the steps in a plant's life cycle.

_____

_____

Summarize

_____

_____

_____

_____

CHAPTER 2

The Animal Kingdom

Vocabulary

 **vertebrate** an animal with a backbone

 **invertebrate** an animal without a backbone

 **endoskeleton** an internal skeleton

 **exoskeleton** a hard covering that protects the bodies of some animals

 **warm-blooded** an animal whose body temperature does not change much

 **cold-blooded** an animal whose body temperature depends on its surroundings

 **skeletal system** the organ system that supports the body

 **muscular system** the organ system, made up of muscles, that moves bones

How are animals different from each other?

nervous system the master control system of the body

respiratory system the organ system that brings oxygen to body cells and removes carbon dioxide

circulatory system the organ system that moves blood through the body

digestive system the organ system that breaks down food

life span how long an organism can be expected to live

metamorphosis a series of changes in a life cycle

clone an offspring that is an exact copy of its parent

heredity the passing of traits from parent to offspring

What are invertebrates?

Animals have many things in common. They also have many things that are different. These features can be used to classify animals.

Symmetry

Animals have different body shapes. Most animals have symmetry (SIM•uh•tree). Symmetry means that parts of an animal's body match up. The body parts match up around a midpoint or line. Some animals have no symmetry. We can use symmetry to classify animals.

▲ A sea urchin has symmetry.

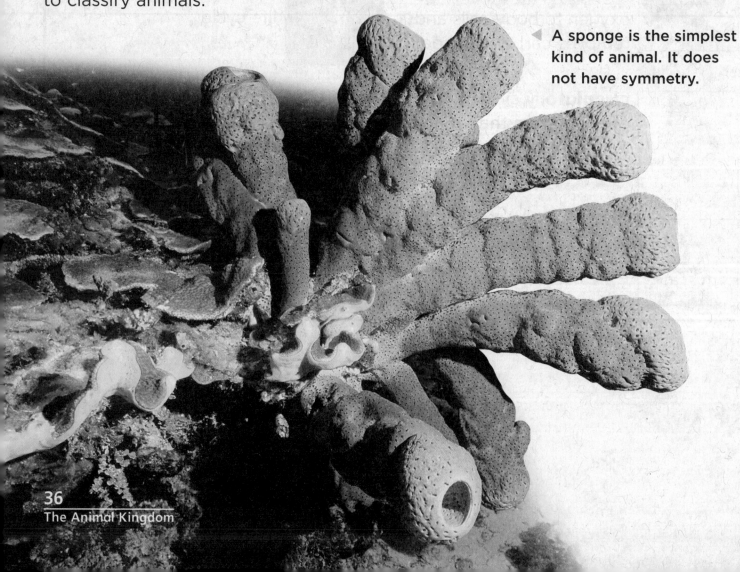

◀ A sponge is the simplest kind of animal. It does not have symmetry.

Backbone or No Backbone

Some animals have a backbone and others do not. A **vertebrate** (VUR•tuh•brayt) is an animal that has a backbone. An animal without a backbone is an **invertebrate** (in•VUR•tuh•brayt).

Most animals are invertebrates. Some invertebrates have a hard outer covering. Others have a skeleton inside their bodies. The picture shows eight groups of invertebrates.

Invertebrate Groups

mollusks

cnidarians

sponges

echinoderms

flatworms

roundworms

segmented worms

arthropods

Read a Photo

How are all of these animals alike?

_____

✔ Quick Check

1. What are two ways we can classify animals?

vertabrates

invertebrates

What are some invertebrates?

	Examples of Invertebrates	
Name	**Facts**	**Examples**
sponges	• simplest kind of invertebrate • most are shaped like a sack with an opening at the top • live under water	• sponge
cnidarians	• have armlike parts called tentacles (TEN•tuh•kuhlz) • stinging cells at the end of each tentacle • stun prey • some stay in one place • others float or swim	• corals • jellyfish
mollusks	• have a shell • some have shells inside the body • most live in water • some swim • others stay in one place • snails and slugs are the only mollusks that live on land	• clams • oysters • octopuses • squid • snails
echinoderms	• spiny skin • have a support structure inside the body called an **endoskeleton** (en•doh•SKEL•i•tuhn)	• sea urchin

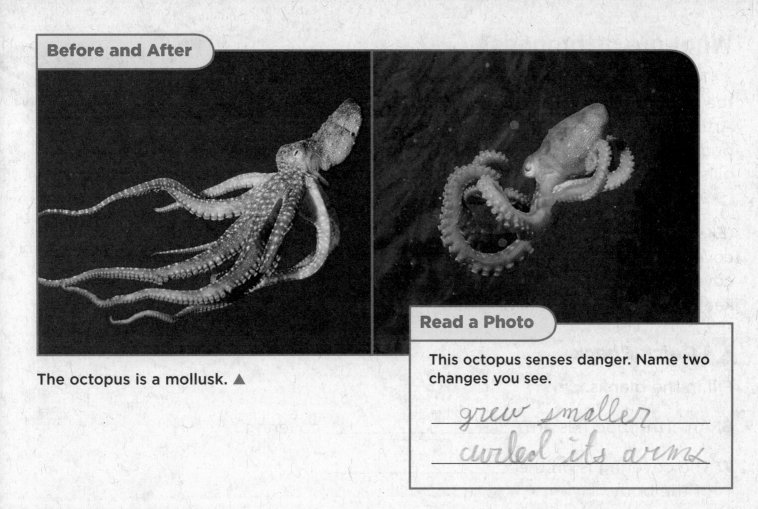

The octopus is a mollusk. ▲

Read a Photo

This octopus senses danger. Name two changes you see.

grew smaller

curled its arms

 **Quick Check**

Tell if each sentence is true or false. If false, correct the sentence.

2. All mollusks live in water. _false_

3. A sponge is the simplest kind of invertebrate.

   true

4. An echinoderm has soft skin. _false_

5. Cnidarians have tentacles. _true_

What are arthropods?

The largest group of invertebrates is the arthropod (AHR•thruh•pod) group. Arthropods have legs that bend. Their bodies have sections. Some breathe with gills. Others breathe through tubes.

Every arthropod has an **exoskeleton** (EK•soh•SKEL•i•tohn). This is a hard covering on the outside of the body. This covering keeps the animal safe. It also keeps the animal from drying out.

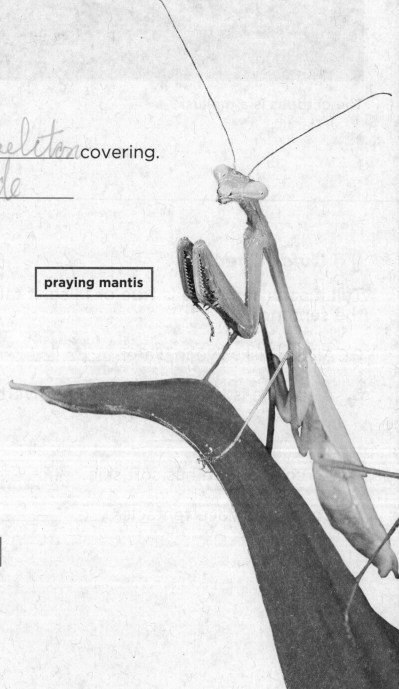

moth

✔ Quick Check

Fill in the blanks.

6. An arthropod has a(n) ___exoskeleton___ covering.

7. The covering is on the ___outside___ of the body.

praying mantis

lady beetles

Insects

weaver ant

Insects have one pair of antennae, three pairs of legs, one or two pairs of wings, and three body sections.

Arachnids

Arachnids (uh•RAK•nidz) include spiders, ticks, and scorpions. They have four pairs of legs, two body sections, and fangs.

huntsman spider

Crustaceans

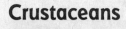

fiddler crab

Crustaceans (kruhs•TAY•shuhnz), such as crabs and shrimps, have two pairs of antennae and two to three body sections. They can chew.

Centipedes and Millipedes

Centipedes have one pair of legs on each body section. A millipede has two pairs of legs on each body section.

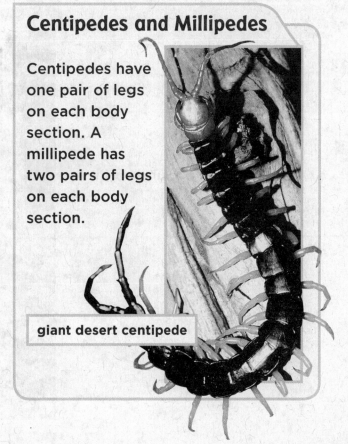

giant desert centipede

How are worms classified?

Three kinds of worms are flatworms, roundworms, and segmented worms.

Flatworms

Flatworms are the simplest worms. Their bodies are flat.

Flatworms have a head and a tail. Most will not hurt other living things. Some live inside other animals.

Roundworms

Roundworms have thin bodies with pointed ends. They are not as thin as flatworms. Food comes into one opening. Waste leaves through another opening. Most roundworms live inside other animals.

▲ A planarian is a flatworm.

A nematode is a roundworm. ▶

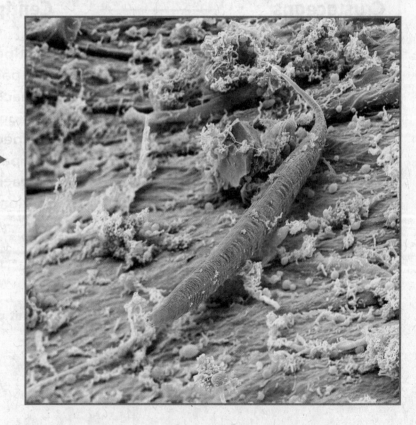

Segmented Worms

The bodies of segmented worms are divided into parts, or segments. The parts are the same, except for the head and tail ends. Each end has an opening for the digestive system.

Most segmented worms live on land. They do not live inside other animals. Earthworms, sandworms, and leeches are segmented worms.

▲ An earthworm is a segmented worm.

✔ Quick Check

Write the letter for the fact about each worm.

8. _c_ flatworms   **a.** thin bodies with pointed ends

9. _a_ roundworms   **b.** do not live in other animals

10. _b_ segmented Worms   **c.** simplest kind of worms

What are vertebrates?

An animal with a backbone is called a vertebrate. The backbone is part of a vertebrate's endoskeleton. It holds up the animal's body. It allows big animals to move around.

There are seven classes of vertebrates: jawless fish, cartilaginous (kahr•tuh•LAJ•un•nuhs) fish, bony fish, amphibians, reptiles, birds, and mammals.

Vertebrates can be warm-blooded. A **warm-blooded** animal has a body temperature that stays the same. It uses energy from food to keep the same body temperature.

Vertebrates can also be cold-blooded. A **cold-blooded** animal has a body temperature that changes. Its body temperature changes with its surroundings. A cold-blooded animal gets heat from outside its body.

◀ **Fish have backbones.**

 **Quick Check**

Fill in the chart.

Types of Vertebrates	
warm-blooded	**11.** Body temperature _doesn't_ change much.
cold-blooded	**12.** Body temperature _does_ with the surroundings.

Classes of Vertebrates

Cold-blooded

cartilaginous fish

reptiles

jawless fish

amphibians

bony fish

Warm-blooded

birds

mammals

Fish

Fish are cold-blooded vertebrates. There are three classes of fish.

Jawless fish and cartilaginous fish both have skeletons made of cartilage (KAHR•tuh•lij). The tip of your nose and your ear are made of cartilage. Sharks are cartilaginous fish.

Jawless fish have a mouth that acts like a suction cup. They have no bones in their mouths. A lamprey is a jawless fish.

Bony fish are the largest class of vertebrates. Their skeletons are made of bones. They are covered in scales. Goldfish, bass, and tuna are bony fish.

✔ Quick Check

13. Which classes of vertebrates are warm-blooded?

birds   mammals

Read a Chart

Are most classes of vertebrates warm-blooded or cold-blooded?

cold-blooded

What are some other vertebrate groups?

Amphibians

Amphibians (am•FIB•ee•uhnz) like frogs are cold-blooded vertebrates. They spend part of their lives in water. They spend the other part on land.

An amphibian's skin needs to stay wet. An adult amphibian has lungs. However, it can also breathe through its skin. If its skin gets too dry, it will die. Amphibians must live near water.

Reptiles

Reptiles (REP•tyelz) include snakes, turtles, and lizards. They are cold-blooded vertebrates. They live on land.

Reptiles have dry skin. Their skin is covered with scales or plates.

Reptiles cannot breathe through their skin. They use lungs. Their eggs have a tough covering. This protects them. It also keeps them from drying out.

lizard

Amphibians and Reptiles

frog

Read a Photo

Which animal needs to stay wet?

amphibians

Birds

Birds are warm-blooded vertebrates with feathers. Their feathers are light but keep birds warm and dry. Birds have beaks and two legs with clawed feet.

Not all birds can fly, but all birds have feathers. Birds have lightweight, hollow bones to help them fly. Strong muscles and lungs also help.

Birds lay eggs with strong shells. Most birds sit on their eggs to keep them warm until they hatch.

Birds are the only animals ▶ that have feathers.

✔ Quick Check

Write one fact about each of the following classes of vertebrates.

14. Amphibians: _can breath through its skin_.

15. Reptiles: _have scaly skin._

16. Birds: _not all can fly._

What are mammals?

You are a mammal. A mammal is a warm-blooded vertebrate with fur or hair. Mammals can live in trees, water, and most other places on Earth!

Most mammals give birth to live young, but some lay eggs. Mammals care for their young. Females make milk to feed their young.

▲ The loris is a mammal with a good sense of sight.

Groups of Mammals

Mammals That Lay Eggs
The only mammals that lay eggs are the duck-billed platypus and the spiny anteater.

Mammals with Pouches
Kangaroos, koalas, and opossums carry their young in pouches until the young are fully grown.

Mammals That Develop Inside
Sheep, bats, apes, and all other mammals develop inside the mother's body.

▲ Both dogs and humans are mammals.

▲ All these animals are mammals, too.

✓ Quick Check

Complete the Main Idea and Details Chart.

Main Idea	Details
Mammals care for their young.	Most mammals give birth to live young.
	Some mammals carry their young in pouches.
	17. *Some give live birth.*

Complete the following sentences.

18. A mammal's body temperature stays the same. It is *warm blooded*.

19. Mammals have fur or *hair*.

20. Mammals can live in trees, *water*, or *other places on Earth*.

How do animals move and sense changes?

Animals have different organ systems. A system is a group of parts that work together.

The Skeletal and Muscular Systems

Bone is living tissue. Bones make up a vertebrate's **skeletal** (SKEL•i•tuhl SIS•tuhm) **system**. The skeletal system is the frame that holds up the animal's body. This frame also protects the organs inside.

The skeletal system works with the muscular system to make the animal move. The **muscular** (MUHS•kyuh•luhr) **system** is made of strong tissues called muscles (MUH•suhlz). When muscles shorten, they pull on bones. When muscles stretch, they push bones. This action makes the body move.

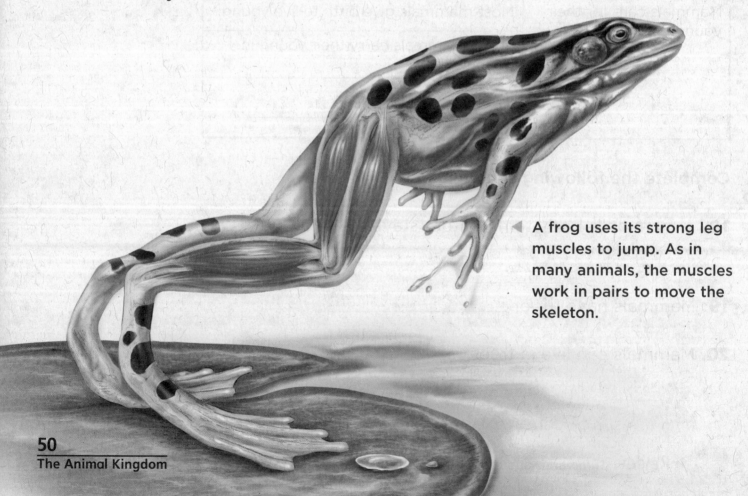

◀ A frog uses its strong leg muscles to jump. As in many animals, the muscles work in pairs to move the skeleton.

The Nervous System

The body is controlled by the **nervous** (NUR•vuhs) **system**. It is made of nerve cells. Invertebrates have simple nervous systems. For example, a sponge has only a few nerve cells. Vertebrates, such as mammals, have millions.

Most animals have a brain and sense organs. The sense organs help animals see, hear, taste, touch, and smell things. The senses help animals know more about the world around them.

▲ Owls have a good sense of sight. Large eyes help them see at night.

✔ Quick Check

Match the organ system to its clue.

21. ___b___ skeletal system   **a.** moves bones

22. ___a___ muscular system   **b.** bones

23. ___c___ nervous system   **c.** controls the body

The dolphin jumps when its brain sends a message to its muscles. ▼

How do gases and blood travel in the body?

The Respiratory System

Animal cells need oxygen. To get that oxygen, most animals have a **respiratory** (RES•pruh•tor•ee) **system**. It brings oxygen to the blood. It also takes carbon dioxide away from the blood. Large animals have gills or lungs. Small animals such as worms do not need gills or lungs. Gases move easily into and out of their bodies.

Adult salamanders have lungs. Like all amphibians, they also breathe through their skin. ▶

▼ Manatees are mammals. They breathe with lungs.

The Circulatory System

The heart, blood, and blood vessels make up the **circulatory** (SUR•kyuh•luh•tor•ee) **system**. Its job is to move blood through the body. The blood carries oxygen, food, and water to the body's cells. In the circulatory system, the heart is the main organ. It pumps lots of blood.

The Excretory System

When cells break down food, they make wastes. The **excretory** (EK•skri•tor•ee) **system** removes the wastes. The liver, kidneys, bladder, skin, and lungs remove wastes.

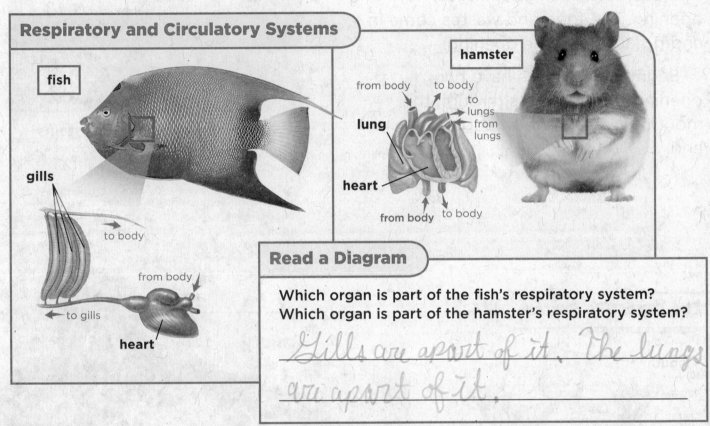

Respiratory and Circulatory Systems

fish

gills

to body

to gills

from body

heart

hamster

from body   to body
to lungs
from lungs

lung

heart

from body   to body

Read a Diagram

Which organ is part of the fish's respiratory system?
Which organ is part of the hamster's respiratory system?

Gills are apart of it. The lungs are apart of it.

✔ Quick Check

Tell if each sentence is true or false. If false, correct the sentence.

24. The respiratory system takes oxygen into a living thing's cells.

False; The respiratory system takes in oxygen into a living things body

25. The brain is the main organ in the circulatory system.

How is food broken down?

Animals take in food for energy. The **digestive** (dye•JES•tiv) **system** helps break down the food. Until food is broken down it cannot be used by cells in the body.

Food and wastes enter and leave through openings in the body. Some invertebrates, such as sponges, have one opening. The food and wastes come in and go out the same opening.

Segmented worms have two openings. Food enters through the mouth. Wastes leave through the tail end.

▲ Sponges have digestive systems with one opening.

The Digestive System

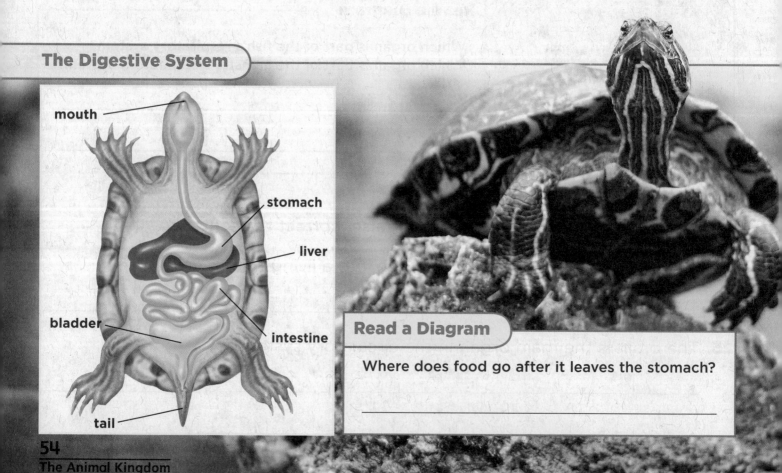

mouth

stomach

liver

bladder

intestine

tail

Read a Diagram

Where does food go after it leaves the stomach?

_____

Reptiles have many parts in their digestive systems. The diagram shows a turtle's digestive system. Mammals have one that is similar. Both reptiles and mammals have many organs in their digestive systems.

Like turtles, snakes have a digestive system with many organs. ▶

◀ A mammal's digestive system is like a reptile's system.

 **Quick Check**

Complete the chart.

Animal	Description of Digestive System
Sponges	26. _____
Segmented worm	27. _____
Reptiles	28. _____
Mammals	29. _____

What are the stages of an animal's life?

Before they reproduce, all living things go through stages of growth and change. The stages of growth, change, and reproduction make up a life cycle. The life cycle of a penguin is shown below.

Life Cycle of a Penguin

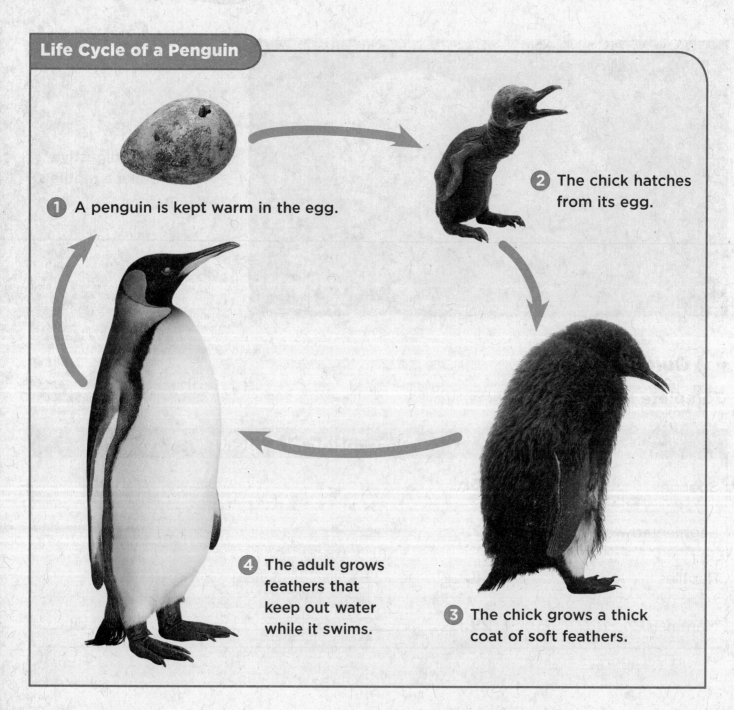

1 A penguin is kept warm in the egg.

2 The chick hatches from its egg.

3 The chick grows a thick coat of soft feathers.

4 The adult grows feathers that keep out water while it swims.

Life Spans

Death is the last stage of an organism's life. An organism's **life span** is how long it usually lives in the wild. Animals have different life spans. Some animals live for days. Others live for many years.

▲ The life span of a skunk is about three years.

A boa constrictor can live as long as 20 years. ▲

A moth has a life span ▶ of about one week.

▲ Koi fish can live to be 100 years old!

 **Quick Check**

Fill in the blanks to complete the sequence of stages in a penguin's life cycle.

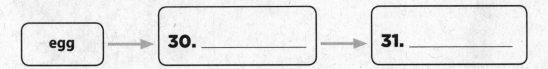

egg → 30. _____ → 31. _____

What is metamorphosis?

Most young animals look like their parents. Other animals grow through metamorphosis (met•uh•MOR•fuh•sis). **Metamorphosis** is a series of very different body forms.

Incomplete Metamorphosis

A damselfly is an insect. It goes though incomplete metamorphosis. That means each body form is not that different from the one before.

As the insect grows and changes, it molts. Molting happens when an animal gets too big for its exoskeleton. It sheds the exoskeleton and grows a new one.

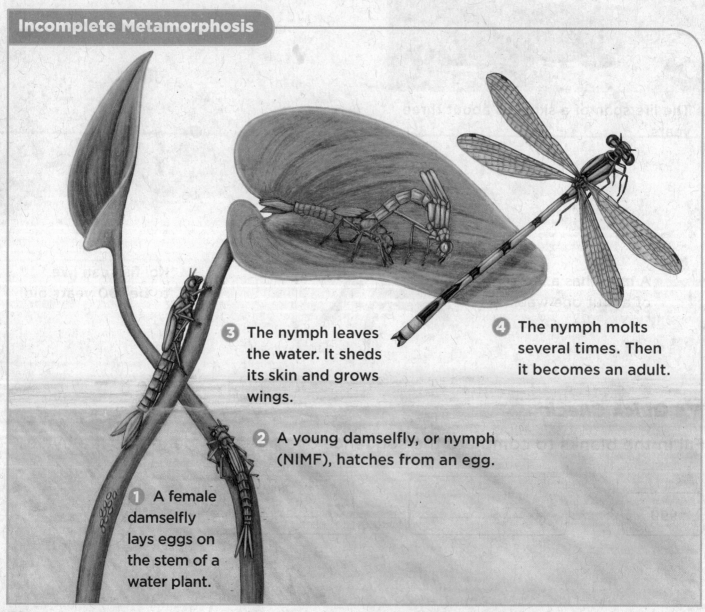

Incomplete Metamorphosis

3 The nymph leaves the water. It sheds its skin and grows wings.

4 The nymph molts several times. Then it becomes an adult.

2 A young damselfly, or nymph (NIMF), hatches from an egg.

1 A female damselfly lays eggs on the stem of a water plant.

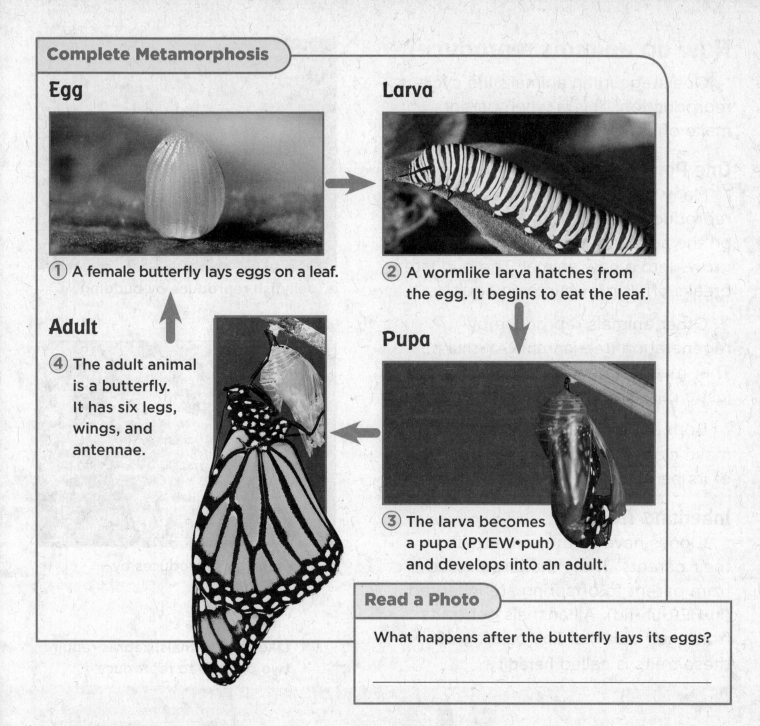

Complete Metamorphosis

Egg

① A female butterfly lays eggs on a leaf.

Larva

② A wormlike larva hatches from the egg. It begins to eat the leaf.

Adult

④ The adult animal is a butterfly. It has six legs, wings, and antennae.

Pupa

③ The larva becomes a pupa (PYEW•puh) and develops into an adult.

Read a Photo

What happens after the butterfly lays its eggs?

_____

Complete Metamorphosis

A butterfly goes through complete metamorphosis. Look at the pictures. Each stage is different.

 **Quick Check**

Complete the sentences.

32. A series of very different body forms is called _____.

How do animals reproduce?

One stage in an animal's life cycle is reproduction. This is when parents make offspring.

One Parent

Many simple invertebrates reproduce by budding. A bud forms on the adult's body. The bud slowly grows into a new animal. The bud later breaks off. It grows into an adult.

Other animals reproduce by regeneration (ree•jen•uh•RAY•shuhn). That means a new animal grows from just a part of the original animal.

Both budding and regeneration make clones. A **clone** is an exact copy of its parent.

Inheriting Traits

Clones have all the same traits as their parents. Traits that are passed from parent to offspring are inherited (in•HER•uh•tid). All animals get traits from their parents. The passing of these traits is called **heredity**.

▲ Jellyfish reproduce by budding.

▲ The sea star reproduces by regeneration.

▼ Like all mammals, zebras require two parents to reproduce.

Two Parents

Another kind of reproduction needs cells from two parents. The female cell is an egg. The male cell is a sperm. Offspring are formed when an egg and a sperm combine. This is called fertilization (fur•tuh•luh•ZAY•shuhn).

As a fertilized egg grows, it becomes more like its parents. It has traits from both parents. It is not exactly like either of them.

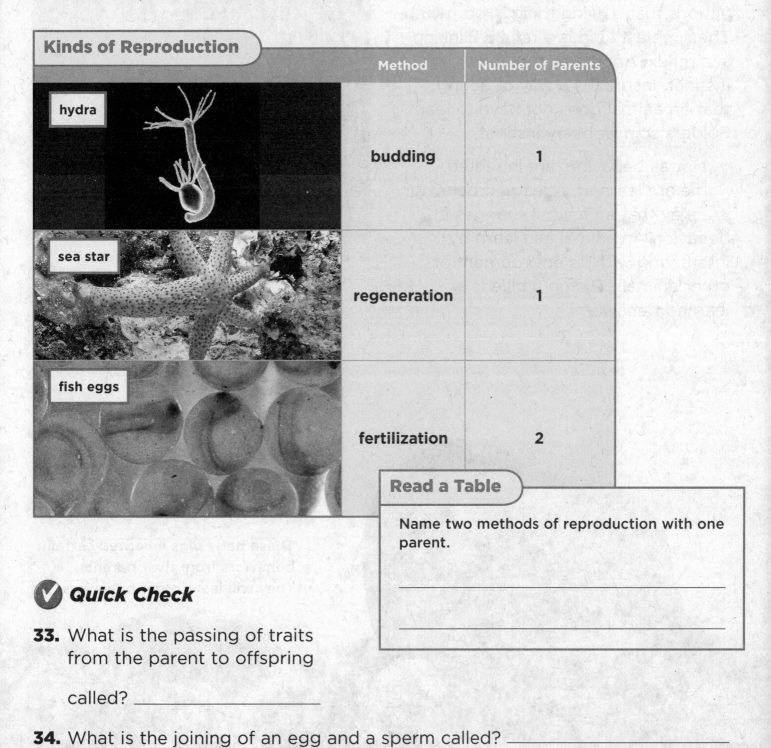

Kinds of Reproduction

	Method	Number of Parents
hydra	budding	1
sea star	regeneration	1
fish eggs	fertilization	2

Read a Table

Name two methods of reproduction with one parent.

_____

_____

✓ Quick Check

33. What is the passing of traits from the parent to offspring called? _____

34. What is the joining of an egg and a sperm called? _____

What is inherited?

Offspring inherit traits from their parents. Some traits, such as eye color, are easy to see. Others are not easy to see.

An inherited behavior is a set of actions that a living thing is born with. The simplest kind is a reflex. Blinking is a reflex. Another example is instinct. Instinct is a way of acting that an animal does not have to learn. Spiders spin webs by instinct.

Not all behaviors are inherited. Some are learned. A learned behavior is a way that an animal changes its behavior. An animal can learn by interacting with its environment or other animals. Riding a bike is a learned behavior.

▲ Birds build their nests by instinct.

These baby pigs inherited certain behaviors from their parents. They will learn other behaviors. ▼

▲ Kayaking is a learned behavior.

✔ Quick Check

Write a definition for each term and give one example of each.

35. learned behavior

_____

_____

_____

36. inherited behavior

_____

_____

_____

The Animal Kingdom

Use the clues below and the words in the box to fill in the crossword puzzle.

clone	invertebrate	respiratory system
cold-blooded	muscular system	skeletal system
exoskeleton	nervous system	warm-blooded

Across

2. a hard covering that protects the body of some animals

4. an animal without a backbone

5. an offspring that is an exact copy of its parent

6. the organ system, made up of muscles, that moves bones

7. the organ system that brings oxygen to body cells and removes carbon dioxide

8. the system that controls the body

9. animals whose temperatures change with their surroundings

Down

1. animals whose body temperatures do not change much

3. the organ system, made up of bones, that supports the body

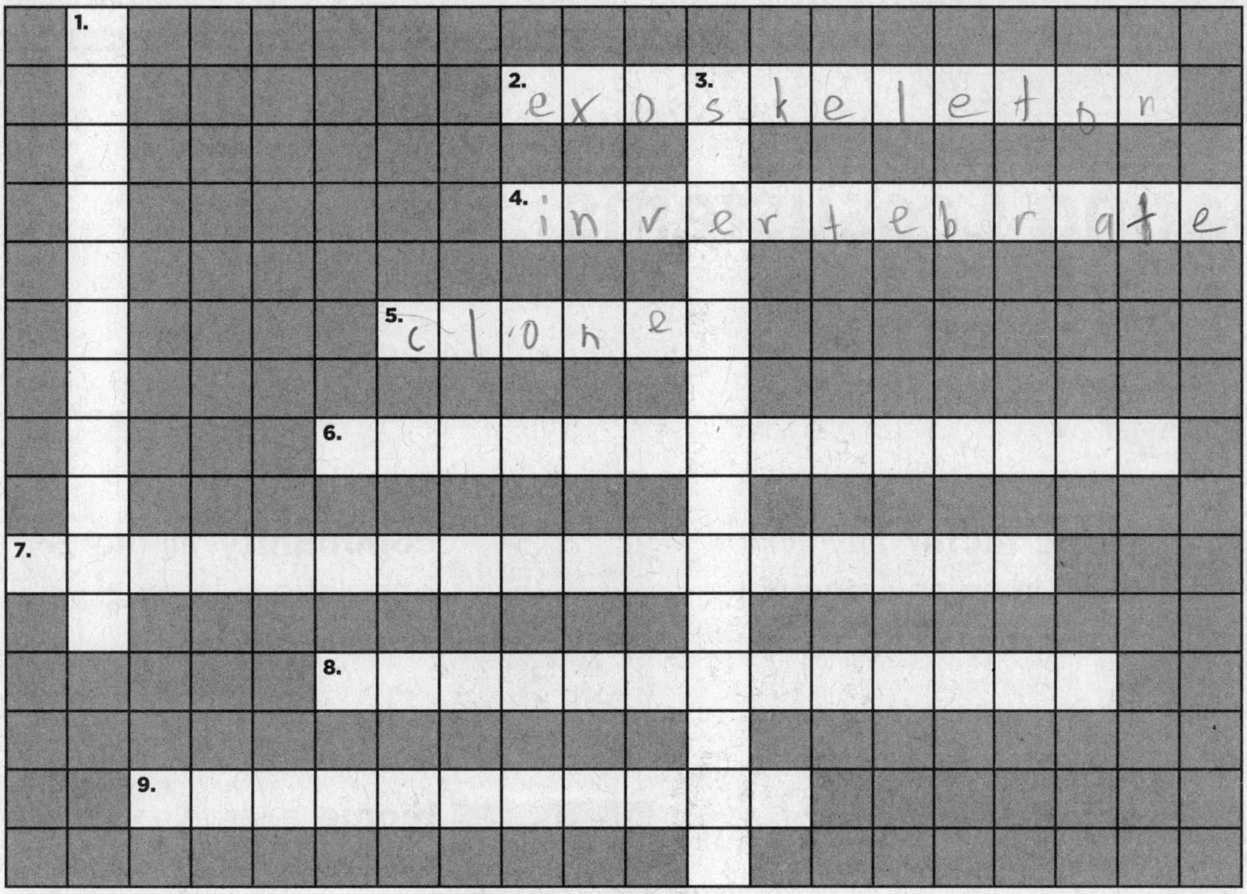
1.
2. exoskeleton
3.
4. invertebrate
5. clone
6.
7.
8.
9.

Summarize

CHAPTER 3

Exploring Ecosystems

Vocabulary

 **biotic factor** any living thing in an environment

 **abiotic factor** any nonliving thing in an environment

 **ecosystem** the interactions among the living and nonliving things in an environment

 **habitat** the place where an organism lives

 **population** all the members of a species that live in the same place

 **community** all the populations in an ecosystem

 **biome** a large ecosystem that has its own kind of climate, soil, and living things

 **grassland** a biome where the main kind of plant is grasses

 **deciduous forest** a forest biome with trees that lose their leaves each year

 **tropical rain forest** a biome that is hot and humid with a lot of rainfall

Where do plants and animals live and how do they depend on each other?

desert a biome with very little rainfall

taiga a cool forest biome in the far north

tundra a cold, dry biome with frozen ground and no trees

producer an organism that makes its own food

consumer an organism that cannot make its own food

decomposer an organism that breaks down the wastes and remains of other living things

food chain the path energy takes in the form of food

food web the food chains that are connected in an ecosystem

energy pyramid the amount of energy at each level of a food chain

What is an ecosystem?

Environments have both living and nonliving things in them. All the living and nonliving things in an environment make up an **ecosystem** (EE•koh•sys•tuhm).

Biotic Factors

All the living things in an environment are called **biotic** (bigh•AH•tik) **factors**. Plants, animals, and bacteria are all biotic factors. You are a biotic factor, too!

Abiotic Factors

Water, rocks, soil, and climate (KLIGH•muht) are **abiotic factors**. Climate is the usual weather pattern in an ecosystem. Abiotic factors are nonliving things.

▲ A hawk is a biotic factor.

◄ Rocks are abiotic factors.

A Pond Ecosystem

Ecosystems

An ecosystem can be as small as a log or as big as a desert. Living things in an ecosystem need the nonliving things. For example, fish need the water in a pond. Living things also need each other. A bird needs plants to make its nest.

Each living thing in an ecosystem has its own place to live. This is called its **habitat**. Different ecosystems have different habitats.

✓ Quick Check

Write the letter of the definition for each word.

1. _____ biotic factor

2. _____ habitat

3. _____ ecosystem

4. _____ abiotic factor

a. the living and nonliving things in an environment

b. a nonliving thing in an environment

c. a living thing in an environment

d. the place where a living thing lives

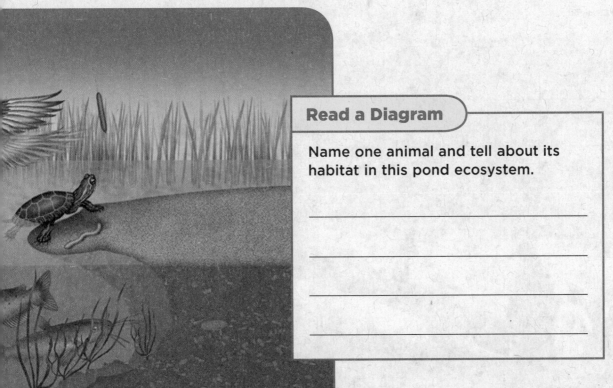

Read a Diagram

Name one animal and tell about its habitat in this pond ecosystem.

_____

_____

_____

_____

What are populations and communities?

A **population** is all the members of a species that live in one place. For example, all the bullfrogs in a pond ecosystem make up one population. The water lilies make up another.

All the populations in an ecosystem make up a **community**. The size of a community depends on important things such as food, shelter, and light. Communities in warm, wet ecosystems tend to be larger than those in cold, dry places.

All the seals in this ecosystem make up one population. All the penguins make up another. The seals and the penguins are part of the same community. ▼

Studying Ecosystems

When scientists want to know about an ecosystem, they look at its populations and communities. A change in a population can affect the community. The reverse is also true.

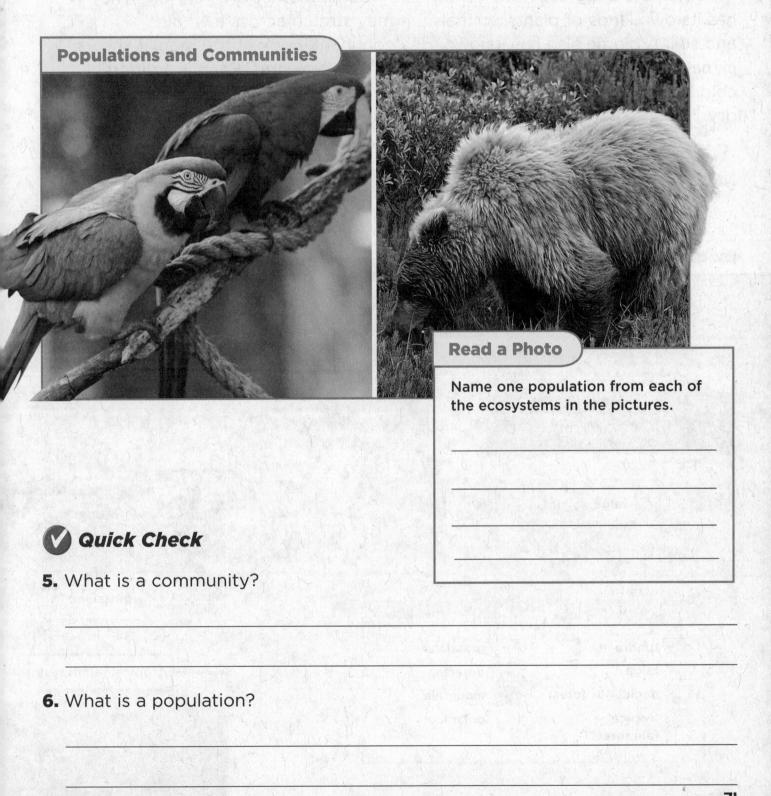

Populations and Communities

Read a Photo

Name one population from each of the ecosystems in the pictures.

_____

_____

_____

_____

✓ Quick Check

5. What is a community?

_____

_____

6. What is a population?

_____

_____

What is a biome?

A **biome** is a big ecosystem that has its own kinds of plants, animals, and soil. A biome also has its own climate. Some biomes are wet and cold. Other biomes are warm and dry.

Some biomes are very big. They may stretch across a whole continent! Look at the map. It shows different biomes. Each is a different color on the map.

Biomes of the World

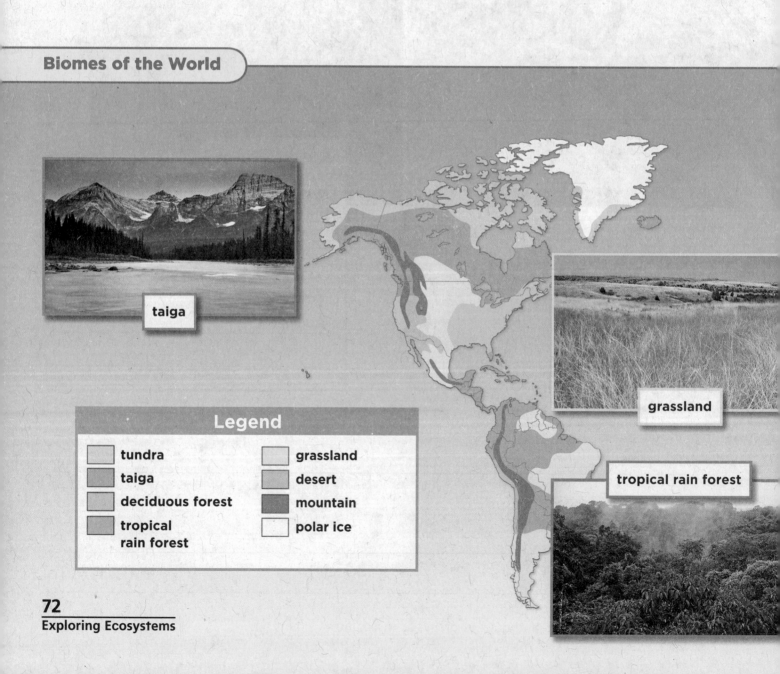

taiga

grassland

tropical rain forest

Legend

tundra	grassland
taiga	desert
deciduous forest	mountain
tropical rain forest	polar ice

The Six Major Biomes

Earth has six major biomes:

- grassland
- deciduous forest
- tropical rain forest
- desert
- taiga
- tundra

 **Quick Check**

List two biomes that you can find in the United States.

7. _____

8. _____

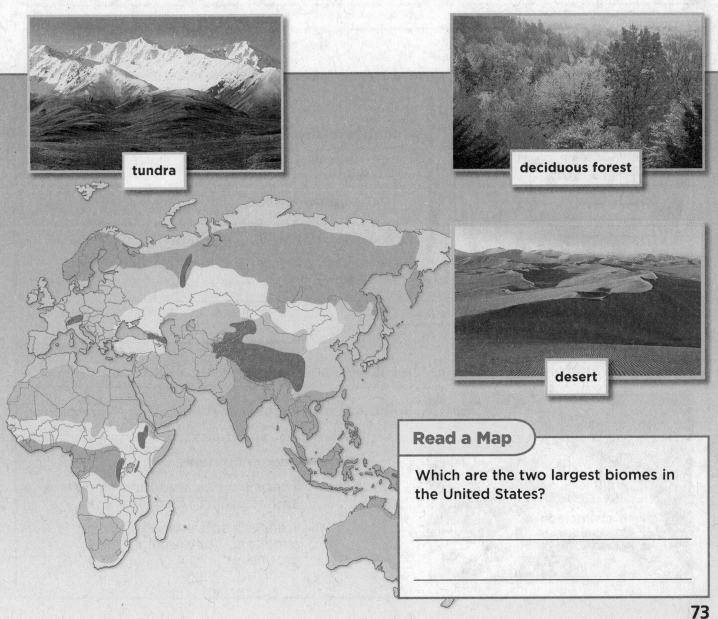

tundra

deciduous forest

desert

Read a Map

Which are the two largest biomes in the United States?

_____

_____

What are grasslands and forests?

Grasslands

A **grassland** is a biome where the main kind of plant is grasses.

bison

Temperature	cool winters, warm to hot summers
Rainfall	moderate
Soils	fertile
Other facts	A prairie is a grassland with a mild climate.
	A savanna is a grassland with shrubs and few trees.

Deciduous Forest

A **deciduous forest** is a biome where many trees lose their leaves each autumn.

raccoon

Temperature	cold to moderate winters, warm summers
Rainfall	year-round
Soils	fertile
Other facts	The ground is usually covered with flowers, ferns, and mosses.

Tropical Rain Forest

A **tropical rain forest** is a biome that is hot and humid with a lot of rainfall.

chameleon

Temperature	warm year-round
Rainfall	wet year-round
Soils	poor in nutrients
Other facts	These biomes are found along and near Earth's equator.
	Trees are so tall and leafy that almost no light reaches the ground.

Grasslands

A grassland biome is mostly grass. The summers there can be hot and dry. Sometimes the grasses burn in the summer. This burning makes the soil good for farming.

Forests

Earth has many kinds of forests. Most forests have three parts:

- The canopy (KAN•uh•pee) is made from the tallest trees. Their leaves spread over the forest like a big umbrella.
- Under the canopy is the understory. It is made of young trees and shrubs. Plants such as orchids live on tree trunks.
- The lowest part is the forest floor. It is so dark that few kinds of plants can grow.

Tree Orchids

Read a Photo

In a rain forest, do orchids live in the canopy, in the understory, or on the forest floor?

_____

 Quick Check

Fill in the table to show how a deciduous forest and a tropical rain forest are different.

	Deciduous Forest	Tropical Rain Forest
9. Temperature in winter		
10. Soils		

What are deserts, taigas, and tundras?

Desert

A **desert** is a barren biome with very little rainfall.

sidewinder

Temperature	very warm during the day and very cold at night
Rainfall	very little
Soils	dry and thin
Other facts	Desert plants and animals must live for a long time with no water.

Taiga

A **taiga** (TY•guh) is a cool forest biome in the far north.

moose elk

Temperature	cold winters, mild summers
Rainfall	moderate
Soils	poor in nutrients
Other facts	The taiga is the largest biome in the world. Conifer trees grow in the taiga. Winters are long and cold. Taiga mammals have thick fur.

Tundra

A **tundra** is a cold, dry biome with frozen ground and no trees

caribou

Temperature	cold, long winters; short, cool summers
Rainfall	some in summer
Soils	frozen
Other facts	Plants grow close to the ground. Many animals hibernate, or go into a deep sleep, for the winter. Some animals go south in the winter to find warmer temperatures.

✓ Quick Check

11. Which biome has the warmest temperatures—desert, taiga, or tundra? _____

Are there water biomes?

Large water ecosystems are not called biomes. This is because water ecosystems have features that are very different from biomes.

Freshwater Ecosystems

Water flows in some freshwater ecosystems. Water flows in streams and rivers. Living things such as fish must be able to live in moving water.

Water stands still in lakes and ponds. It is easier for algae and plants to grow there.

Wetlands and Estuaries

Plants grow from under the water in a wetland ecosystem. Salt marshes and mangrove swamps are wetlands. An estuary is where ocean and freshwater ecosystems meet. Both are rich in plant and animal life.

Ocean Ecosystems

Scientists classify an ocean habitat by how deep it is. Algae are the main producers in the ocean. They live near the surface to get sunlight. Some fish live deep in the ocean. Others live near the shore.

▲ Alligators live in swamps and other wetlands.

▲ It is always cold and dark in the deep ocean. Unusual fish live there.

✔ Quick Check

Name four water ecosystems.

12._____

13._____

14._____

15._____

How do organisms depend on one another?

Scientists sort organisms by what they eat. Organisms within a community need each other to survive. They count on each other for food.

Producers

Producers are organisms that make their own food. Most producers make food using energy from the Sun. Green plants are producers on land. Algae are the main producers in water. Every ecosystem must have producers.

▼ **Green plants are producers on land.**

Consumers

Consumers are organisms that cannot make their own food. Animals are consumers. They get their energy by eating other living things.

There are different kinds of consumers:

- Herbivores (UR•buh•vorz) eat only plants. Rabbits and deer are herbivores.
- Carnivores (KAHR•nuh•vorz) eat other animals. Eagles, tigers, and sharks are carnivores.
- Omnivores (OM•nuh•vorz) eat both plants and other animals. Opossums, raccoons, and bears are omnivores.

Decomposers

Decomposers break down the wastes and remains of other organisms. They return nutrients to the ecosystem. Bacteria and fungi are decomposers. They are recyclers.

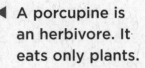
◄ A porcupine is an herbivore. It eats only plants.

An opossum is an ► omnivore. It eats plants and animals.

◄ A falcon is a carnivore. It eats only meat.

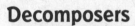

✔ Quick Check

Complete the Main Idea and Details chart below.

Main Idea	Details
Organisms within an ecosystem rely on each other.	Producers make their own food using sunlight.
	16. Consumers
	17. Decomposers

What is a food chain?

Energy passes from one organism to another in a **food chain**. The energy starts with the Sun. The sun's energy is stored in the food producers make. Then the energy moves from producers to consumers. Decomposers get energy from producers and consumers.

A Pond Food Chain

This diagram shows a food chain. Each arrow points away from the organism that is eaten. The arrow shows the direction energy moves.

Algae and green plants start a pond food chain. An insect such as the mayfly eats algae. The insect uses some energy from the algae to live. It may store some of the energy, as well.

A consumer such as this sunfish eats the mayfly. A blue heron eats the sunfish. Each organism uses some energy from its food. Each organism may also store some of the energy.

After the plants and animals die, they become food for decomposers. They break down the remains into nutrients. Plants use the nutrients from the soil to grow.

Pond Food Chain

blue heron

sunfish

mayfly

algae

Read a Diagram

Which consumer eats the mayfly?

_____

_____

LOG ON *Science in Motion* Watch decomposers in action at www.macmillanmh.com

A Land Food Chain

A food chain on land is similar to one in a pond. The land food chain starts with grasses, trees, or other producers.

The thistle is the producer in this food chain. The caterpillar is an herbivore. It eats the leaves of the thistle. The mantis eats the caterpillar. The skink eats the mantis. The owl eats the skink.

Decomposers are not in the food chain diagram. However, they are an important part of every food chain.

✓ Quick Check

18. Which organism is the producer in the pond food chain?

_____

19. Which organisms in the land food chain will become food for decomposers?

_____

_____

_____

_____

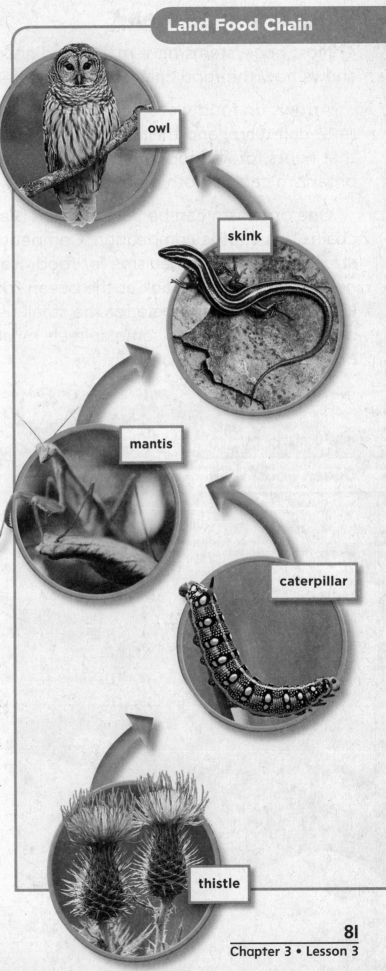

Land Food Chain

owl

skink

mantis

caterpillar

thistle

What is a food web?

Most ecosystems have many food chains. A **food web** shows how the food chains are connected.

Arrows on food webs connect predators (PRE•duh•tuhrz) and their prey. A predator is a carnivore that hunts for its food. What it hunts is its prey. An organism can be both predator and prey.

One organism can be part of more than one food chain. This causes competition. Competition is the struggle between organisms for food, water, and other needs. For example, look at the ocean food web. Three kinds of animals compete for the small fish. If there are not enough small fish, some animals must find new food or move away.

Ocean Food Web

Read a Diagram

Identify one predator and its prey in this food web.

_____

_____

What is an energy pyramid?

An **energy pyramid** shows how much energy is at each level of a food chain.

The bottom layer always shows the producers. The second level shows herbivores. Each herbivore gets energy the producers have stored. However, each herbivore gets only $\frac{1}{10}$ of the producer's energy.

The top levels show omnivores and carnivores. Each level has fewer organisms than the last. Only about $\frac{1}{10}$ of the energy gets passed from one level to the next. There is little energy left at the top of the pyramid.

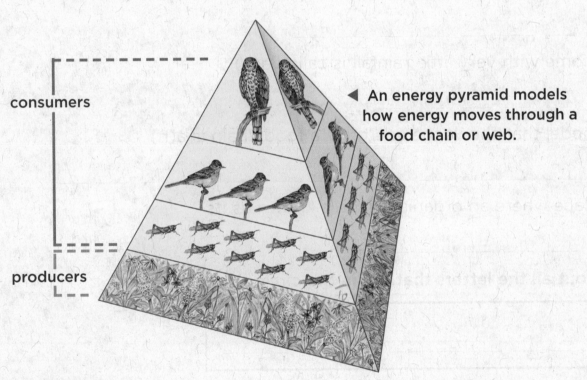

consumers

producers

◀ An energy pyramid models how energy moves through a food chain or web.

✓ Quick Check

20. A model that shows the amount of energy at each level of a food chain is

called a(n) _____.

Exploring Ecosystems

Complete the sentences below. Fill each blank with one letter.

1. A biome where the main kind of plant is grasses is a(n)

g r a s s l a n (d)

2. A living thing in an environment is a(n)

b i (a) t i c   f a c t (e) r

3. An organism that cannot make its own food is a(n)

c o n s u (m)(e) r

4. A biome with very little rainfall is called a(n)

d e (s) e (r) t

5. A model that shows the energy in a food chain is a(n)

(e) n e r g y (p) y r a m i d

6. A place where an organism lives is known as its

h a b i t a t

Write out all the letters that are in the circles.

d a a me s r e p

**Use the letters from inside the circles above to name
the word described below.**

7. Which organism is not usually shown in a food chain
diagram?

decomposer

Use the clues below to fill in the crossword puzzle.

Across

2. an animal that is hunted by other animals

3. an organism that makes its own food

4. a cool forest biome in the far north

5. the struggle between organisms for food, water, and other needs

6. a carnivore that hunts for its food

7. a cold, dry biome with frozen ground and no trees

Down

1. the interactions among the living and nonliving things in an environment

3. all the members of a species that live in the same place

Summarize

_____

_____

_____

CHAPTER 4

Surviving in Ecosystems

Vocabulary

adaptation a trait that helps a living thing survive in its environment

mimicry when one kind of organism looks like another kind of organism

hibernate to rest or sleep through the cold winter

stimulus something in an environment that causes a living thing to respond

camouflage a color or pattern that helps an animal blend in with its surroundings

Why do plants and animals live in different places and what happens when those places change?

tropism a plant's response to a stimulus

endangered a living thing that has few of its kind left

accommodation an organism's response to changes in its ecosystem

extinct when there are no members of a species left alive

What are adaptations?

It can be hard to live in an environment. Adaptations (a•dap•TAY•shunz) help living things live in certain environments.

Adaptations help living things survive in their environment. To survive is to continue to live. Adaptations can be traits such as a fish's gills. A bird's beak is also an adaptation.

Some adaptations are traits such as body parts. Others are behaviors. Some adaptations help organisms move. Some help them get food. Others help them live in certain climates.

Some adaptations help an animal defend itself. A porcupine has quills. The quills protect it from predators.

A camel's hump stores fat for times when food is scarce. ▼

Desert Adaptations

A desert environment is dry. Desert animals have adaptations that save water.

A desert bird has feathers that hold water. This helps it carry water to its young in the nest. A kangaroo rat gets water from food. It never needs to drink.

Animals in hot deserts have adaptations for staying cool. A fennec fox has large ears. The large ears get rid of heat.

Camels have many adaptations for life in a desert. They store fat in humps. They can close their nostrils. This keeps sand out. They have wide hoofs. This helps them walk on sand.

The kangaroo rat gets water from the seeds it eats. ▶

✔ Quick Check

1. What helps living things survive in their environment?

_____

_____

Animal Adaptations

Read a Photo

Which fox's ears help it live in a hot desert?

_____

What are some other adaptations of animals?

Animals have different adaptations depending on where they live. Animals in cold environments may migrate (MIGH•grayt). To migrate is to change location from time to time. Birds migrate from cooler to warmer areas. There are many other kinds of adaptations.

Some animals that live in cold environments **hibernate** (HYE•buhr•nayt). To hibernate is to rest or sleep through the cold winter. The photo shows a marmot. Marmots hibernate in family groups. These animals live off their body fat and use very little energy.

Camouflage (KAM•uh•flahzh) is a color or pattern that lets an animal blend in with its surroundings. Camouflage helps prey hide from predators. The picture shows a leopard. Camouflage also helps predators sneak up on prey.

king snake

coral snake

This harmless king snake looks like a poisonous coral snake. Predators may be fooled and stay away from the king snake. When one kind of organism looks like another kind of organism, it is called **mimicry** (MIM•i•kree).

Read a Photo

How do the fish help the turtle survive?

_____

_____

How else do animals survive?

Some animals depend on other animals. For example, some animals make their homes on other animals. A flea might live in a mammal's fur. The flea gets food and shelter. This interaction only helps the flea. The other animal is hurt.

Other interactions between animals can help them both. Sometimes algae covers the shells of sea turtles. Fish eat the algae off the shells. The fish get food. The turtles get a clean shell. The interaction helps both the fish and the turtles.

✓ Quick Check

Match the type of interaction to its explanation.

2. _____ interaction   **a.** A king snake looks like a coral snake.

3. _____ hibernation   **b.** A bear sleeps through winter.

4. _____ mimicry   **c.** A reef fish eats algae off of a sea turtle.

How do plants respond to their environment?

A **stimulus** (STIM•yuh•luhs) is something that causes a living thing to respond. A change in an environment can be a stimulus. Plants can respond to these changes.

Stimulus and Response

A plant responds to a stimulus in the way it grows. Sunlight is a stimulus. Plants respond by growing toward it. Gravity is another stimulus. Its pull is the reason that plant roots grow down into the Earth. Water is a stimulus for plants, too. Roots will grow toward a water source.

Tropism

The responses of plants to light, water, and gravity are tropisms (TROH•pi–zumz). A **tropism** is a plant's response to a stimulus. Plants also show tropisms to chemicals and heat.

✔ Quick Check

5. What is the name for the response of a plant to a stimulus?

_____

_____

Tropism Experiment

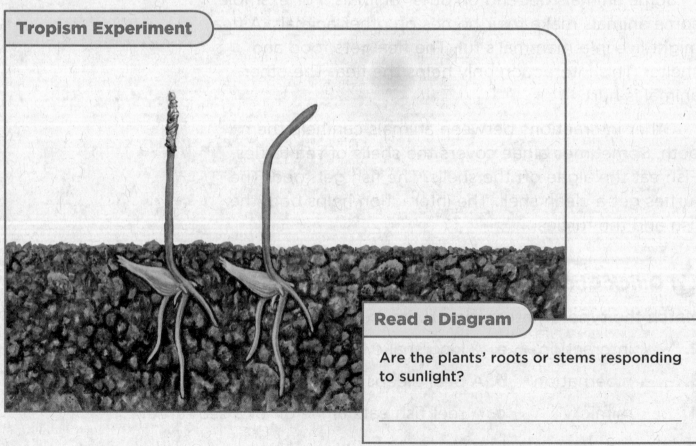

Read a Diagram

Are the plants' roots or stems responding to sunlight?

_____

Desert Adaptations

Read a Photo

Which part of the cactus helps keep water inside the plant?

_____

_____

Science in Motion See adaptations of other desert plants at www.macmillanmh.com

What are some plant adaptations?

Plants have adaptations for different environments. For example, desert plants need to keep from losing water.

A cactus is a desert plant. It holds water like a sponge. It has a thick, waxy cover. This keeps the water inside.

Plants in other environments have different adaptations. Cold winter air can harm the leaves of trees. Some trees lose their leaves in winter. This adaptation protects the leaves from being damaged.

Without leaves, the tree cannot make food. Instead it uses stored food. In spring, the tree grows new leaves.

✓ Quick Check

6. How does a cactus keep from losing all of its water?

_____

What causes an ecosystem to change?

Ecosystems are always changing. Some changes make it hard for living things to survive.

Natural Events

Change in ecosystems is part of nature. Hurricanes may destroy coastal wetlands. Too much rain may cause a landslide. Too little rain can cause a drought (DROWT). In a drought, soil dries up.

It can take many years for an ecosystem to recover from these kinds of changes. In 1980, a volcano called Mount Saint Helens erupted in Washington. The ash and lava killed nearby plants and trees. However, over time, the ash made the soil rich in nutrients. Now the ecosystem is healthier than before.

Natural Change in Ecosystems

Mount Saint Helens in 1980

Mount Saint Helens in 1995

Read a Photo

How did Mount Saint Helens change between 1980 and 1995?

_____

Living Things

Living things can change ecosystems, too. For example, giant groups of locusts can gather to search for food. These groups can have 50 million locusts in them! They eat all the plants in their path. They can leave a whole community without food.

locust

Some living things help an ecosystem. In a wetland ecosystem, an alligator may use its body to dig a hole. Slowly, the hole fills with water.

Alligators use the "gator hole" to survive droughts. Other wetland animals, such as birds, use the gator hole, too. They find food, water, and shelter there.

▲ Gator holes help many animals survive times of drought.

✓ Quick Check

Fill in the missing cause or effect in each row of the diagram.

Cause	→	Effect
7. _____	→	valley full of ash
hurricane	→	8. _____
too much rain	→	9. _____
10. _____	→	soil dries up

How do people change ecosystems?

People change ecosystems. Some changes harm the ecosystem. Others help it.

Deforestation

People cut down trees in forests. They use the trees to build things. When too many trees are cut down, it is called deforestation (dee•for•uh•STAY•shun). Deforestation destroys forest habitats for many organisms. Living things lose their homes and sources of food.

Overpopulation

People use resources. Some, such as water and space, may become hard to find. When too many people live in an area, it is called overpopulation (oh•ver•pop•yew•LAY•shun). Plants and animals can also be overpopulated. Too many living things can use up all the resources in an ecosystem.

Pollution

Adding things that can hurt air, water, or land is pollution (puh•LEW•shun). Litter is a kind of pollution. Pollution can kill living things in an ecosystem.

Manufactured Ecosystems

an airplane being sunk

concrete "reef balls" on the sand

Protection

People may hurt ecosystems, but they can also help. You can help by:

• recycling paper, glass, and plastic
• turning off the water when you brush your teeth

✓ Quick Check

What are three ways people hurt an ecosystem?

11. _____

12. _____

13. _____

▲ The green liquid flowing out of this drainage pipe is pollution.

a subway car being sunk

7871

7788

Read a Photo

How can putting these things in the water help underwater ecosystems?

_____

What happens when ecosystems change?

When an ecosystem changes, organisms need to change. They may move to a new ecosystem. If they do not change or move, they will die.

Accommodating

An organism's response to changes in its ecosystem is called an **accommodation** (uh•kom•uh•DAY•shun). An animal may need to change its behaviors and habits. For example, a fire may destroy all fresh plants in an ecosystem. Some herbivores will eat tree bark instead of fresh plants.

Moving Away

Not all animals can adjust to changes. These animals must find a new place that meets their needs. Their search may take them far away.

Extinction

If an organism does not meet its needs, it will die. A living thing that has few of its kind left is **endangered** (en•DAYN•jurd). Some endangered plants and animals can become extinct (ek•STINGKT). A living thing is **extinct** when there are no more of its kind alive.

▲ When few plants are around, a deer may eat the bark of a tree.

 Quick Check

Complete each sentence with one of these words.

accommodation   endangered   extinct

14. When there are no animals of one kind left, that animal is _____.

15. An organism's response to change is a(n)

_____.

16. If a living thing has few of its kind left, it is

_____.

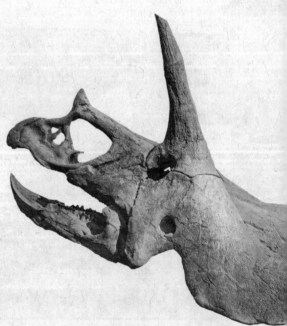

▲ Dinosaurs are extinct.

How can people prevent extinction?

People want to stop endangered species from becoming extinct. For example, scientists are working to save giant pandas. Long ago there were many giant pandas in China. Today, pandas are endangered. To help save the panda, some land in China has been set aside.

Now, panda cubs are born in safe places. Many other endangered species, such as tigers, are protected. Countries have laws to protect them. Scientists hope this will prevent extinctions.

✔️ *Quick Check*

17. Scientists are trying to save pandas because they

are _____.

The giant panda is an endangered species. ▼

Surviving in Ecosystems

Choose the letter of the best answer.

1. A snake looks like another kind of snake that predators leave alone. This is an example of

   a. hibernation.

   b. mimicry.

   c. adaptation.

   d. accommodation.

2. An individual organism's response to changes in its ecosystem is a(n)

   a. adaptation.

   b. camouflage.

   c. accommodation.

   d. endangered.

3. A plant growing to sunlight is an example of

   a. tropism.

   b. mimicry.

   c. gravity.

   d. hibernation.

4. A fox in a cold climate has thick fur. This is a(n)

   a. extinction.

   b. hibernation.

   c. stimulus.

   d. adaptation.

5. When there are no more of a kind of living thing left alive, it is

   a. endangered.

   b. camouflage.

   c. stimulus.

   d. extinct.

6. An insect is the same color as the plant it lives on. This is an example of

   a. camouflage.

   b. tropism.

   c. mimicry.

   d. hibernation.

Match the definition with the vocabulary word.

1. ____ an adaptation in which one kind of living thing looks like another

2. ____ a trait that helps a living thing survive in its environment

3. ____ a color or pattern that helps an animal blend in with its surroundings

4. ____ the reaction of a plant to something in its environment

5. ____ something in an environment that causes a living thing to respond

6. ____ to rest or sleep through the cold winter

7. ____ a living thing that has few of its kind left

a. adaptation

b. camouflage

c. endangered

d. hibernate

e. mimicry

f. stimulus

g. tropism

Summarize

_____

_____

_____

_____

_____

_____

_____

CHAPTER 5

Shaping Earth

Vocabulary

 **crust** solid rock that makes up Earth's outermost layer

 **mantle** the layer of solid rock below Earth's crust

 **outer core** the liquid layer below Earth's mantle

 **inner core** a sphere of solid material at Earth's center

 **mountain** a tall landform that rises to a peak

 **earthquake** a sudden shaking of Earth's crust

 **seismic wave** a vibration caused by an earthquake

 **seismograph** a tool that finds and records earthquakes

 **volcano** a mountain of once molten rock formed around an opening in Earth's crust

What causes Earth's surface to change?

weathering a slow process that breaks rocks into smaller pieces

erosion the weathering and removal of rock or soil

deposition the dropping off of weathered rock

flood a flow of water over land that is normally dry

tornado a column of spinning wind that moves across the ground in a narrow path

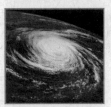
hurricane a very large, swirling storm

landslide the sudden downhill movement of rock and soil

avalanche the sudden downhill movement of ice and snow

What does Earth's land look like?

Earth's land is not flat. There are deep cuts in its surface. Tall hills rise above nearby land. Earth's surface is made of different natural features. They are called landforms.

Tallest and Flattest

The tallest landforms are mountains. They rise to a peak at the top. Some mountains are volcanoes formed by melted rock.

The flattest landforms are plains. Plains are large areas of land with no hills or mountains.

Landforms Shaped by Water

Flowing water can shape the land. Fast rivers can cut deep valleys. In some places, they can make steep-sided valleys called canyons. Ocean waves shape land, too. They can make a flat beach or a rocky cliff.

Landforms Shaped by Wind

Wind can pile sand into large hills in deserts and on beaches. These hills are called sand dunes.

▲ From outer space Earth looks like a smooth, mostly blue ball.

✔ Quick Check

Write the letter of the landform for each description.

1. _____ steep-sided valley

2. _____ tallest landform

3. _____ flat

4. _____ hills on a beach

a. mountain
b. sand dunes
c. plain
d. canyon

Landforms of the United States

Read a Map

Name each of the numbered landforms surrounding the map.

_____

_____

_____

_____

What does it look like where water meets land?

Water always flows downhill. Moving water can carry bits of rock and soil. Different landforms are shaped by water.

Deltas

As a river flows into an ocean, it slows down and stops. It drops off bits of rock it carried. The bits of rock form a delta. A delta is the landform where a river meets the ocean.

Drainage Basins

Rivers flow downhill in valleys. Small rivers may empty into larger rivers. The area of land drained by a river is called a drainage (DRAY•nihj) basin. When rivers join together they form larger drainage basins.

▲ This delta formed where a river entered the sea.

From Land to Sea

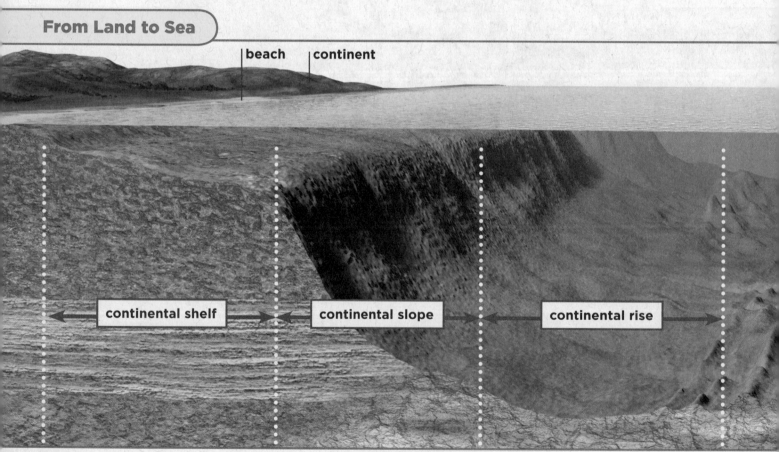

beach | continent

continental shelf

continental slope

continental rise

Beyond the Continent

The continental shelf is the land below the ocean. The continental shelf is a part of the continent. It is covered by ocean. It can stretch into the sea for miles.

The continental shelf starts on the shore. It ends at a feature called the continental slope. This feature is a steep part of the continent. It slopes toward the ocean floor.

At the base of the slope is the continental rise. The rise connects the continent to the ocean floor.

Most of the ocean floor is flat. However, long mountain ranges can be found along the bottom of some oceans. These are called ocean ridges.

ocean ridge

✓ Quick Check

5. What is a delta?

_____

_____

_____

_____

Read a Diagram

Which landform lies just beyond the continental shelf in an ocean?

_____

What is below Earth's surface?

Look at the diagram. It shows Earth's surface. It also shows the layers below Earth's surface. Earth has four layers.

1 **Crust** is the solid rock that makes up Earth's outermost layer. It can crack. Landforms and underwater features are found on the crust.

2 **Mantle** is the layer of rock below Earth's crust. High temperature or pressure can change its shape. It can flow.

3 **Outer core** is the liquid layer below Earth's mantle. It is made mostly of melted iron.

4 **Inner core** is the sphere of solid material at Earth's center. It is the hottest part of Earth. It is probably made of iron.

▲ Earth's crust can have many different shapes. Wind caused these features.

✔ Quick Check

6. Describe Earth's crust.

_____

_____

7. Describe Earth's core.

_____

_____

How does Earth's crust move?

Earth's crust sits on the mantle. When the mantle moves, so does the crust.

Plates

Earth's crust is made of huge plates of rocks. These plates sit on the mantle. Both crust and mantle are solid. However, the mantle can flow when it gets hot. When the mantle flows, the plates on top move, too.

Earth's plates move slowly. They move about as fast as your fingernails grow. Plates are being pulled apart in some places. In other places plates crash into one another. The crust changes where plates pull apart. It also changes when plates crash together.

Mountains in the Making

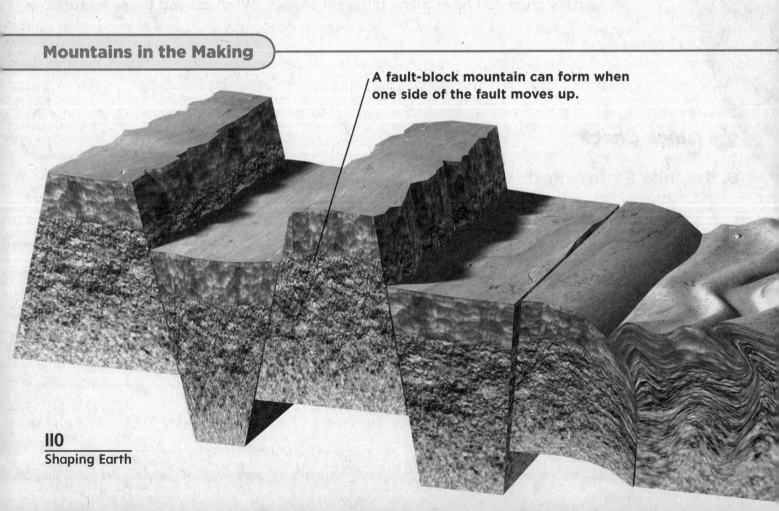

A fault-block mountain can form when one side of the fault moves up.

Faults

The moving mantle pushes, pulls, and twists the plates. This creates cracks in Earth's crust. A long, narrow crack in the crust is a called a fault.

Rocks on one side of a fault can slide up while the other slips down. The side that is lifted may form a mountain. If the ground is lifted over a wide area, a plateau (pla•TOH) may form. A plateau is a high landform with a flat top.

Folds

Some plates push into each other. The land between them is squeezed. A fold forms on Earth's surface. A fold is a bend in the rock layers. If the plates keep pushing, a fold becomes a mountain. A **mountain** is a tall landform that rises to a peak.

fault-block mountain

fold mountain

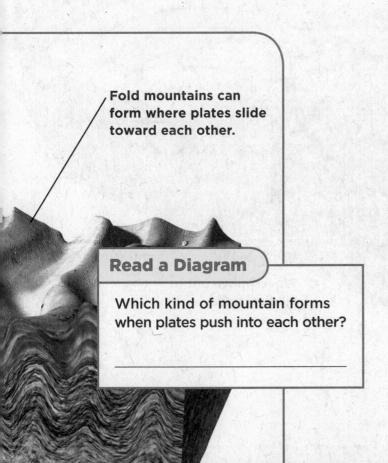

Fold mountains can form where plates slide toward each other.

Read a Diagram

Which kind of mountain forms when plates push into each other?

_____

✔ *Quick Check*

Write the name of the landform for each description.

8. A long, narrow crack in the crust is a(n) _____.

9. A bend in the rock layers is a(n) _____.

What causes earthquakes?

An **earthquake** is a sudden shaking of Earth's crust. Crust moving along a fault causes earthquakes. An earthquake begins under the ground. The sudden movement along the fault gives off energy that shakes Earth's crust. Waves of energy go out through the crust in all directions.

Earthquake Safety

Most earthquakes cannot be felt. However, in a major earthquake, buildings and roads may break apart. Indoors, you can stay safe by ducking under a table or in a doorway. Keep away from walls and windows. Outdoors, stay away from trees or anything that might fall. Do not go near power lines.

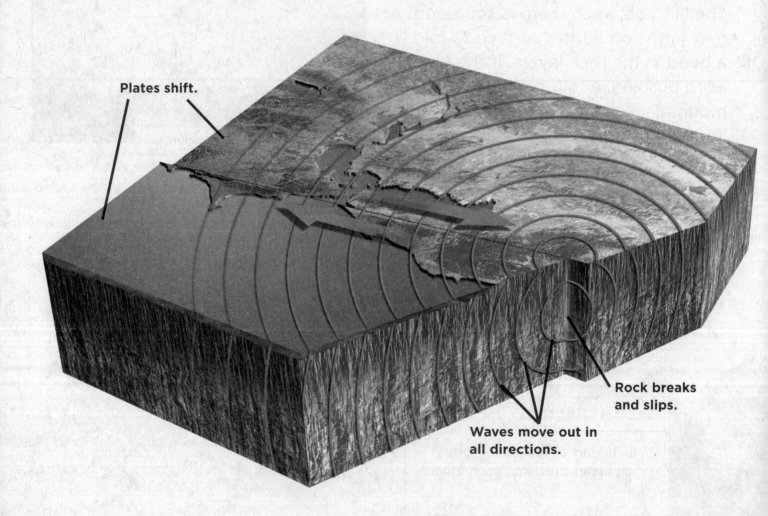

Plates shift.

Rock breaks and slips.

Waves move out in all directions.

This building was damaged during an earthquake in San Francisco. ▼

Earthquakes in the Ocean

If an earthquake is strong enough, it can suddenly lift part of the ocean floor. A giant ocean wave, or tsunami (soo•NAH•mee), may head toward the shore. A tsunami can destroy everything in its path.

 **Quick Check**

10. What is an earthquake?

_____

_____

How do scientists study earthquakes?

Seismic waves are the vibrations caused by earthquakes. Scientists learn about earthquakes by studying seismic waves.

Measuring Seismic Waves

Scientists use a tool called a seismograph (SIZE•muh•graf) to measure how strong an earthquake is. The **seismograph** finds and records earthquakes. It measures seismic waves. It shows the waves as curvy lines. The lines show the strength of the earthquake.

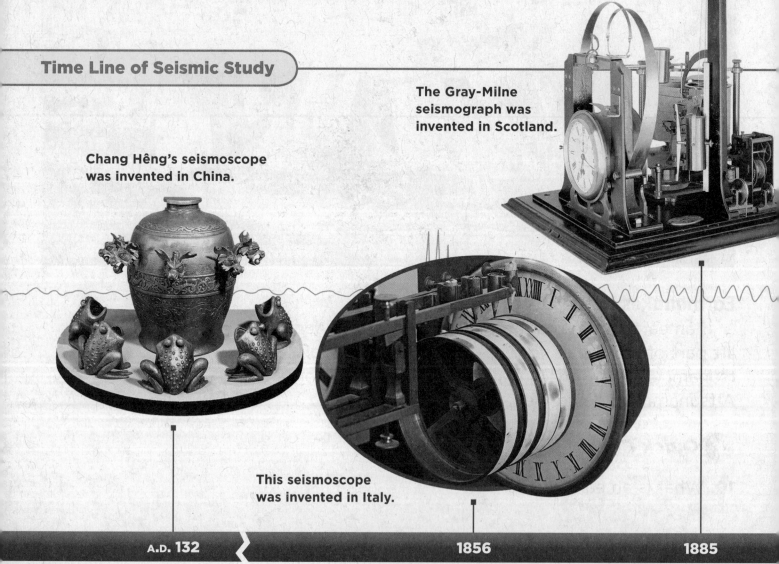

Time Line of Seismic Study

The Gray-Milne seismograph was invented in Scotland.

Chang Hêng's seismoscope was invented in China.

This seismoscope was invented in Italy.

| A.D. 132 | 1856 | 1885 |

Seismic Networks

Earthquake scientists have seismographs around the world. Information comes in from each station. Scientists use this information to learn where an earthquake started. They also find out how deep inside Earth it started.

✔ Quick Check

11. Earthquakes cause vibrations called

_____.

12. To measure how strong an earthquake is, scientists

use a(n) _____.

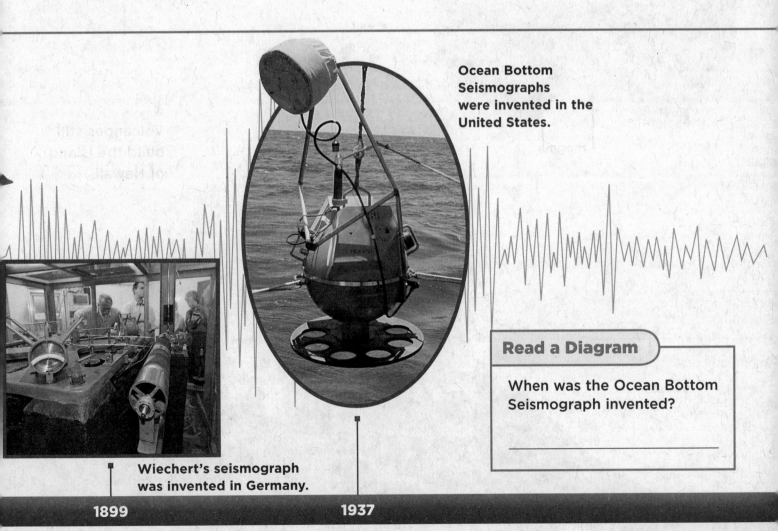

Ocean Bottom Seismographs were invented in the United States.

Wiechert's seismograph was invented in Germany.

1899

1937

Read a Diagram

When was the Ocean Bottom Seismograph invented?

_____

What is a volcano?

A **volcano** is a mountain of once-melted rock that forms around an opening in Earth's crust. Sometimes molten material from inside Earth is forced out an opening. A volcanic eruption can send melted rock, gases, ash, or small rocks into the air. Melted rock is called magma. Magma is called lava when it reaches Earth's surface. A volcano can make a large mountain if it erupts often.

magma

◀ Volcanoes still build the Island of Hawaii.

Where Volcanoes Form

Most volcanoes form at the edges of plates. As one plate sinks under another, some of it begins to melt. A volcano forms if melted rock rises to the surface. Volcanoes also form where Earth's plates move apart. The space between the plates lets magma rise to Earth's surface.

Hot spots are places where plates move over a very hot part of the mantle. Volcanoes can form at hot spots, too.

Lava is melted rock that cools at Earth's surface.

✓ Quick Check

13. What is a hot spot?

_____

What is weathering?

Rocks break apart over time. They change shape.
Weathering is the slow process that breaks rocks into
smaller pieces. Some causes of weathering are flowing
water, rain, and wind.

Physical Weathering

Rocks can change size and shape without changing
what they are made of. This process is called physical
weathering. Living things can cause physical weathering.
For example, a plant's roots can break a rock apart.

Water causes physical weathering in several ways:
- Moving water in rivers moves sharp rocks,
  making them round and smooth.
- Waves can break off small pieces of rock.
- Rain may get into small cracks in a rock
  and freeze. Freezing water widens the
  cracks.

**Weathering made this
boulder break apart.** ▼

▲ Chemical weathering can form limestone caves such as this one.

Chemical Weathering

Chemical weathering changes what a rock is made of. Oxygen, acids, and carbon dioxide all cause chemical weathering. For example, water and oxygen can weather iron. They change iron into rust.

Some rocks can also rust. Rocks that have iron in them can rust. Water and carbon dioxide can dissolve limestone. That is how limestone caves form.

✓ **Quick Check**

14. What is physical weathering?

_____

_____

_____

15. What is chemical weathering?

_____

_____

_____

What is erosion?

The weathering and removal of rock and soil is called **erosion**. Weathering and erosion shape the land.

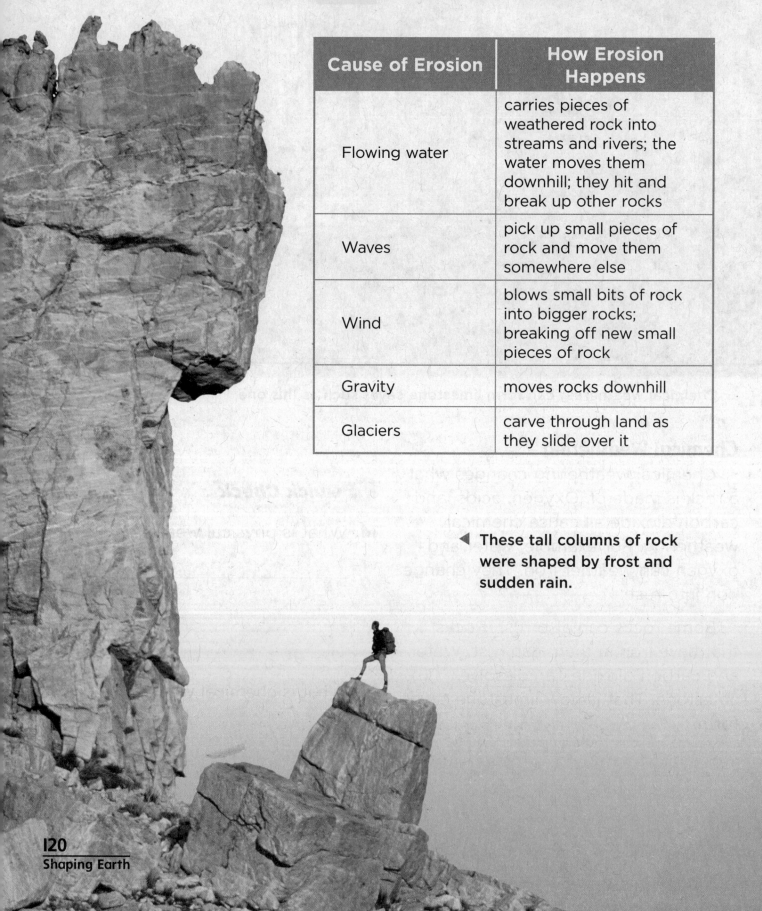

Cause of Erosion	How Erosion Happens
Flowing water	carries pieces of weathered rock into streams and rivers; the water moves them downhill; they hit and break up other rocks
Waves	pick up small pieces of rock and move them somewhere else
Wind	blows small bits of rock into bigger rocks; breaking off new small pieces of rock
Gravity	moves rocks downhill
Glaciers	carve through land as they slide over it

◀ These tall columns of rock were shaped by frost and sudden rain.

Rivers Erode the Land

Rivers and streams weather rock and pick up soil as they flow. A powerful river can carve out the land beneath it. Some of the rocks and soil are dropped off on the banks. Some get carried to the mouth of the river.

Deposition

Deposition is the dropping off of weathered rock. Deposition by water builds deltas, riverbanks, and beaches. Deposition by wind forms sand dunes.

Rivers Shape the Land

Read a Photo

What formed this canyon?

_____

✔ Quick Check

16. The weathering and moving of rock is called

_____.

17. The dropping off of weathered rock is called

_____.

How do glaciers shape the land?

Glaciers (GLAY•shuhrz) are thick, moving sheets of ice. They form in very cold places. Snow makes the glacier heavier. The heavy glacier will begin to flow downhill. As it moves, it drags rocks and flattens a path. A glacier can make a valley wider and steeper.

▲ A glacier carved this valley in Alaska.

A Glacier Deposits Land

moraine

glacial till

moraine

What Glaciers Leave Behind

As glaciers melt, they leave behind the rocks they carried. These are called glacial debris (GLAY•shuhl duh•BREE). The glacier drops most of this debris at its end, or terminus.

Glacial debris is made of gravel, rocks, sand, and clay. The glacial debris can form hills. These hills are called moraines (muh•RAYNZ).

 **Quick Check**

18. What is a glacier?

_____

_____

_____

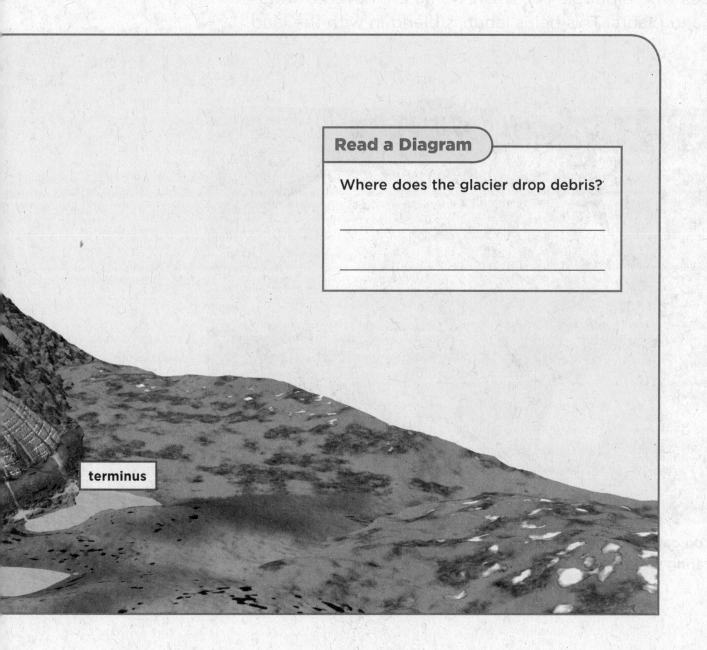

Read a Diagram

Where does the glacier drop debris?

_____

_____

terminus

How do people shape the land?

Weathering is a slow process. People can change Earth's land much more quickly.

One way people change the land is by mining it. Mining is digging into the land. People look for useful materials such as minerals, metals, or fuel. People cut down forests to clear the land. They use the trees to make products.

People need places to put their trash. Landfills are places where people pile trash. Some are covered with soil and plants. This helps landfills blend in with the land.

▲ You can help the land by planting and caring for trees.

▲ Part of this evergreen forest has been cut down.

✓ Quick Check

Describe three ways people can change Earth.

19._____

_____

20._____

_____

21._____

_____

How do floods and fires change the land?

Floods and fires change the land quickly.

Floods

Heavy rains or melting snow mean more water than usual. The flowing water can spill out of rivers or streams. The ground may not soak up all the water. Then it floods (FLUHDZ) the land. A **flood** is a flow of water onto land that is usually dry. City streets flood when water is not carried away fast enough.

Floods can also help nature. After a flood, new soil is put on the land. The nutrients in the soil help plants grow.

Fires

When there is little rain, trees in forests are very dry. Many fires start when lightning hits dry wood. A fire can burn down a forest. Animals lose their habitats. Most places recover from natural fires.

✔ Quick Check

22. How do floods help nature?

_____

_____

23. What can cause a fire?

_____

_____

Read a Photo

Name one change the fire made.

_____

Name one change the flood made.

_____

How do storms change the land?

Tornadoes and hurricanes are severe storms. They can change the land. They cause damage to the land.

Tornadoes

A thunderstorm can grow into a dangerous storm called a tornado (tor•NAY•doh). **Tornadoes** are columns of spinning wind. They move across the ground in a narrow path. A tornado destroys everything in its path.

a tornado

Hurricanes

A **hurricane** is a very large, swirling storm. The eye, or center, of the hurricane is calm. However, walls of wind, clouds, and rains spiral out from the eye. A hurricane is much bigger than a tornado. Its swirling arms can span hundreds of kilometers.

Hurricanes form over warm oceans near the equator. Their winds cause large waves as they travel. The waves of water may be pushed onto shore by the winds. Huge floods may happen. Heavy rains add to the floods. A hurricane can pull trees out of the ground and destroy buildings. It can change an ecosystem in one day.

 Quick Check

Fill in the blanks in this cause-and-effect chart.

Cause	→	Effect
24. _____	→	a column of spinning wind that destroys everything in its path
25. _____	→	storm that forms over water and can change an ecosystem in one day

Hurricane Damage

Read a Photo

What kind of damage did the hurricane do to the town?

_____

_____

How do landslides change the land?

After a heavy rain, loose rock and soil can slide down a slope. A **landslide** is the sudden downhill movement of a lot of rock and soil.

Gravity pulls on rock and soil. Gravity causes rocks and other objects to fall from high places to low places. Gravity can move a house down a slope.

Avalanches

An **avalanche** (AV•uh•lanch) is like a landslide. It is the sudden downhill movement of a lot of ice and snow.

Scientists try to know when and where landslides and avalanches will happen. It is hard to know when one will start.

✔ **Quick Check**

Where do landslides and avalanches happen?

26. _____

_____

An avalanche such as this one can shift the snowy landscape. These people are in trouble! ▼

Shaping Earth

Match the words in the first column to the best answer in the second column.

1. _e_ hurricane

2. _f_ seismograph

3. _d_ crust

4. _a_ tornado

5. _b_ erosion

6. _h_ mantle

7. _c_ earthquake

8. _g_ weathering

a. a column of spinning wind that moves across the ground in a narrow path

b. the weathering and removal of rock and soil

c. a sudden shaking of Earth's crust

d. rock that makes up Earth's outermost layer

e. a very large, swirling storm

f. a tool that finds and records earthquakes

g. a slow process that breaks rocks into smaller pieces

h. the layer of melted rock below Earth's crust

Answer the following questions. Use words from the first column above.

9. Which of Earth's layers is shaken in an earthquake?

_____

10. What is a strong storm that begins over warm water?

_____

Read each clue, then fill in the crossword puzzle.

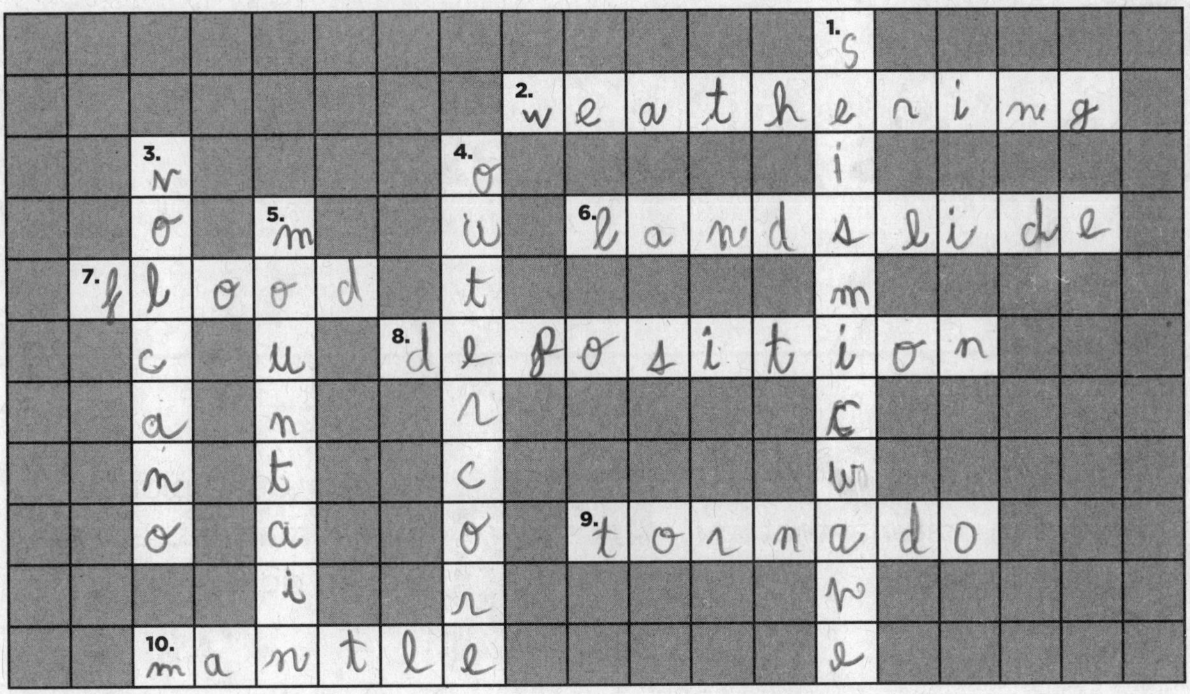

1. seismicwave
2. weathering
3. volcano
4. outercore
5. mountain
6. landslide
7. flood
8. deposition
9. tornado
10. mantle

Across

2. can be physical or chemical
6. the sudden downhill movement of a lot of rock and soil
7. a great flow of water over land that is normally dry
8. dropping off of weathered rock
9. a dangerous storm that begins as a thunderstorm
10. the layer of Earth that makes plates move

Down

1. a vibration caused by an earthquake
3. a mountain that forms around an opening in Earth's crust
4. the liquid layer below Earth's mantle
5. a tall landform that rises to a peak

Summarize

CHAPTER 6

Saving Earth's Resources

Vocabulary

 **mineral** a natural, nonliving material that makes up rock

 **igneous rock** rock formed when melted rock cools

 **sedimentary rock** rock formed from sediments that are stuck together

 **metamorphic rock** rock formed from another kind of rock through heat and pressure

 **rock cycle** the process in which rocks change from one type to another

 **humus** dead plant or animal matter that has broken down

 **horizon** a layer of soil

 **pore space** the space between particles of soil

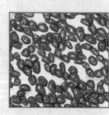 **permeability** tells how fast water goes through a material

 **fossil** the remains of an organism that lived long ago

What are Earth's resources and how can we conserve them?

 **fossil fuel** an energy source that formed millions of years ago

 **nonrenewable resource** a useful material that cannot be replaced easily

 **renewable resource** a useful material that can be replaced quickly

 **reservoir** a storage area for fresh water

 **pollution** harmful or unwanted material that has been added to the environment

 **conservation** using resources wisely

 **reduce** to use less of something

 **reuse** to use something over again

 **recycle** to make a new product from old materials

What is a mineral?

Why do rocks look different from each other? They are made of different minerals (MIN•uh•ruhlz). A **mineral** is a natural, nonliving substance. Minerals are the building blocks of rocks.

Scientists identify minerals by looking at their properties. A property is a characteristic.

Color

One property is color. Talc is white. Topaz can be blue. However, you cannot identify a mineral just by color. Many minerals have the same color. Other minerals can come in many different colors.

Hardness

Hardness is the ability to scratch or be scratched by another mineral. A soft mineral is easy to scratch. A hard mineral is hard to scratch. The Mohs' scale uses numbers to tell how hard or soft a mineral is. Each mineral has a number from one to ten. Ten is the hardest kind of mineral.

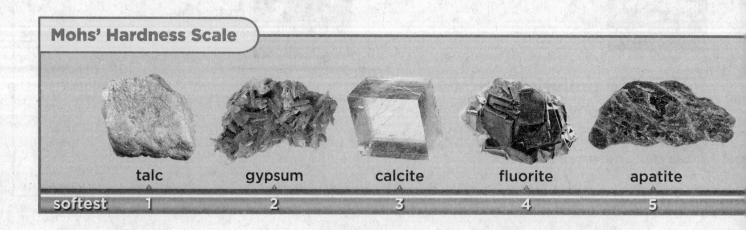

Mohs' Hardness Scale

talc	gypsum	calcite	fluorite	apatite
softest 1	2	3	4	5

Properties of Minerals

Mineral	mica	pyrite	feldspar	hematite
Color	white, green, silver, or brown	gold or brassy yellow	white, pink, gray, or smoky black	gray or brown
Luster	pearly	metallic	dull or glassy	metallic or dull
Streak	white	green-black	white	red
Hardness	2–2.5	6–6.5	6–6.5	5–6

Luster

Luster is the way light bounces off a mineral. Some minerals have a shiny luster. They are metallic. Other minerals are dull or glassy.

Streak

When you scratch a mineral along a white tile, it leaves behind a mark. Streak is the color of that mark. The streak may be different than the color of the mineral's surface.

✔ Quick Check

1. What does the Mohs' scale measure?

_____

_____

Read a Table

What is the hardest mineral?

_____

feldspar	quartz	topaz	corundum	diamond
6	7	8	9	10 hardest

What are igneous and sedimentary rocks?

Igneous rocks and sedimentary rocks are two different kinds of rocks. They form in different ways.

Igneous Rocks

Melted rock below Earth's surface is called magma. Above Earth's surface, melted rock is called lava. Melted rock can cool and harden. Rocks that form from melted rock are **igneous** (IG•nee•uhs) **rocks**.

The size of a rock's mineral grains is related to how fast it cools. An igneous rock that cools slowly has large mineral grains. An igneous rock that cools quickly has small mineral grains.

▲ Long ago, North American hunters used obsidian to make spearheads.

Examples of Igneous Rocks

- Obsidian (uhb•SID•ee•uhn) is smooth and shiny. It forms when lava cools quickly.
- Basalt (buh•SAWLT) has small mineral grains. It also forms from cooling lava.
- Granite (GRA•nit) forms underground slowly. It has large mineral grains.

The "steps" of Giant's Causeway in Ireland are made of basalt. ▼

basalt

Sedimentary Rocks

Sandstone is made of tiny pieces called sediment (SED•uh•muhnt). Some sediment is from rocks or minerals. Some is bits of plants, shells, or other materials.

Sedimentary rocks form from sediment that is pressed and stuck together.

1. Wind and water break up rocks. This makes sediment.

2. Sediment piles up in one place and forms layers.

3. The heavy top layers squeeze out water and air from layers below.

4. Dissolved minerals cement the sediment together.

▲ This sandstone is made up of layers of different sediments.

Relative Age

Sedimentary rocks often have layers. The layers are stacked by their relative ages. Relative age is the age of one thing compared to another. The top layers of a rock are young compared to the bottom layers.

 **Quick Check**

Fill in the missing effects in the chart.

Cause	⟶	Effect
Magma cools and hardens	⟶	**2.** _____
Sediment is pressed together	⟶	**3.** _____

What are metamorphic rocks?

When rocks are under high heat and pressure, their properties can change. Rocks changed by high heat and pressure are called **metamorphic** (met•uh•MOR•fik) **rocks**. They can form from igneous, sedimentary, or other metamorphic rocks.

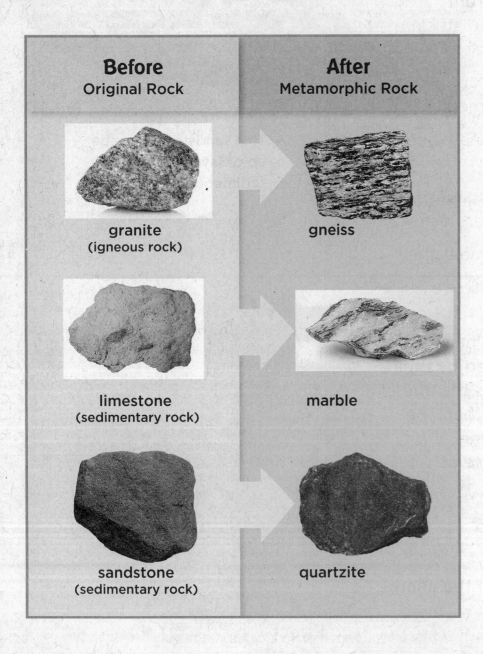

Before Original Rock	**After** Metamorphic Rock
granite (igneous rock)	gneiss
limestone (sedimentary rock)	marble
sandstone (sedimentary rock)	quartzite

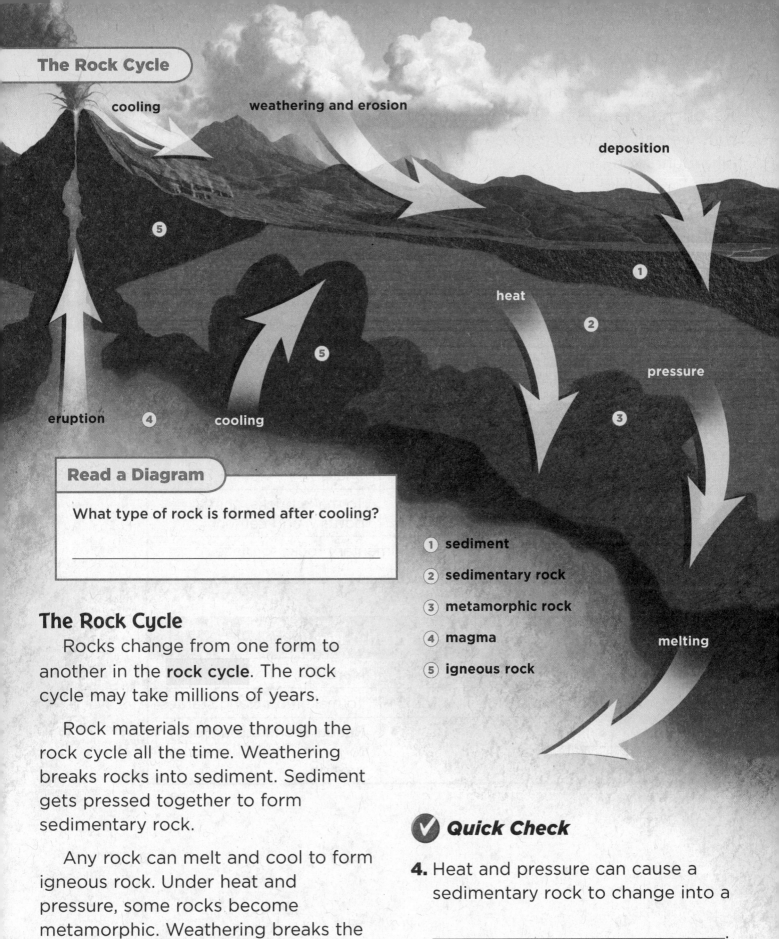

The Rock Cycle

cooling

weathering and erosion

deposition

⑤

heat

②

pressure

eruption ④ cooling

⑤

③

① sediment
② sedimentary rock
③ metamorphic rock
④ magma
⑤ igneous rock

melting

Read a Diagram

What type of rock is formed after cooling?

_____

The Rock Cycle

Rocks change from one form to another in the **rock cycle**. The rock cycle may take millions of years.

Rock materials move through the rock cycle all the time. Weathering breaks rocks into sediment. Sediment gets pressed together to form sedimentary rock.

Any rock can melt and cool to form igneous rock. Under heat and pressure, some rocks become metamorphic. Weathering breaks the rocks apart. The cycle goes on.

✓ *Quick Check*

4. Heat and pressure can cause a sedimentary rock to change into a

_____.

How do we use rocks?

Rocks and minerals are resources. Resources are materials from Earth that we can use. You can see examples all around you.

Type	Igneous Rocks	
Example	Granite	Pumice
Uses	To build schools and other structures	To make soaps and cleansers
Why	Strong and long-lasting	Rough texture helps to scrub off dirt.

Type	Sedimentary Rocks	
Example	Limestone	Shale
Uses	To make glass and cement	To make bricks, china, pottery, and cement
Fun fact	Scientists use layers of sedimentary rocks to piece together Earth's history.	

Type	Metamorphic Rocks	
Example	Slate	Marble
Uses	To make shingles for roofs	To make floors and statues
Why	Waterproof	Beautiful and strong, easy to carve, resists fire and erosion

Saving Earth's Resources

Quartzite is used to make glass.

This guardian lion in Thailand is made of marble. ▼

✓ **Quick Check**

Name two ways we use rocks.

5. _____

_____

6. _____

_____

What is soil made of?

If you look closely at soil, you find many different things. You may see small pieces of rocks and minerals. You may also see humus (HYEW•muhs). **Humus** is nonliving plant or animal matter. Soil also has water, air, and living things.

How Soil Forms

Soil can take hundreds of years to form, or even longer. Pieces of rock get smaller and smaller through weathering. Plants begin to grow in these sediments. Their roots push the sediments deeper. Animals can move these pieces and mix them around.

When plants and animals die, their bodies break down. Humus forms. Humus has nutrients for new plants to grow.

Weathering Caused by Living Things

Read a Photo

What did this rabbit do to the soil?

_____

_____

LOG ON *Science in Motion* Watch animals in the soil at www.macmillanmh.com

Soil Horizons

Soil forms in layers called **horizons** (huh•RY•zuhnz). Each horizon has a different amount of rock and humus. A soil profile shows these horizons. In some places, the soil profile might look like the one on this page.

The A horizon is the layer of soil at the surface. It is also called topsoil. It is rich in humus and minerals. The B horizon is usually lighter and harder than topsoil. The C horizon is found below the B horizon. Underneath the C horizon is bedrock. The C horizon is made of weathered bedrock.

The rock and humus that make up soil are not the same in all places. That is why soil profiles are different from place to place.

A soil profile shows the layers in soil. ▶

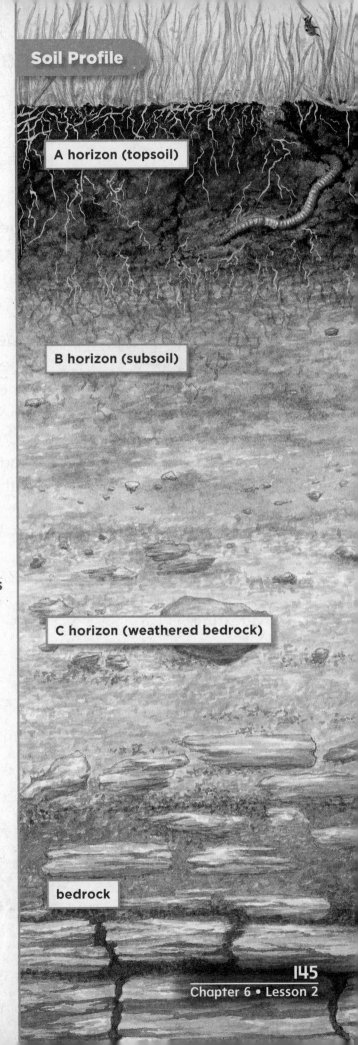

Soil Profile

A horizon (topsoil)

B horizon (subsoil)

C horizon (weathered bedrock)

bedrock

✓ Quick Check

7. What does a soil profile show?

_____

_____

8. Why are soil profiles different from place to place?

important

_____

_____

▲ Sandy soil has a coarse texture.

▲ Silty soil has a medium texture.

▲ Clay soil has a fine texture.

What are some properties of soil?

There are many types of soil. Each type of soil has its own properties. One soil property is color. Another is texture. Soil texture is the size of the pieces of soil.

Pore Spaces

The spaces between pieces of soil are called **pore spaces**. The pore spaces in soil act like filters. They remove some things from water as the water moves through. This keeps the water clean.

Permeability

The sizes and numbers of pore spaces affect a soil's **permeability** (pur•mee•uh•BIL•i•tee). Permeability describes how fast water goes through a material. Sandy soil has high permeability. Its particles are packed loosely together. They hold little water.

Permeability of Soils

Read a Photo

Which soil has the lower permeability—fine soil or coarse soil?

_____

fine soil

cattail

Factors That Affect Soils

The properties of soils depend on:

- climate
- ecosystem
- bedrock
- steepness of the land
- time

Steep slopes erode quickly. They often have thin soils. Thin soils are not good for growing crops. Thicker soils can build up on flat lands.

Water, wind, and ice can move soil. Minerals in moved soil may be very different from those in the bedrock.

 Quick Check

Write the meaning of each term.

9. pore spaces _____

10. permeability _____

mallee tree

coarse soil

Why is soil type important?

Topsoil is home to many living things. Soil type is important to these living things. All living things need at least a little water. Many living things also need air. Plants and animals can live in soil only if the soil can hold both water and air.

Some soils hold very little water. Other soils hold too much water. Farmers grow crops in soils that hold the right amount of water for their crops.

Farmers depend on soil to grow crops.

Soil and Crops

Sandy soil is not good for growing crops. It holds too little water. This causes crops to dry up. Water moves through sandy soil quickly. The water carries away nutrients.

Fine soil is also not good for growing most crops. It has low permeability. Water may stay in the pore spaces for a long time. A crop can die from too much water.

✓ Quick Check

11. Why is topsoil important?

_____

_____

_____

12. Which kind of soil lets water move through quickly?

_____

_____

_____

13. Which soil is good for many crops?

_____

_____

_____

Plant Adaptations to Soils

Desert plants are adapted to grow in sandy soils.

Medium-textured soils are good for many crops.

Some kinds of grapes grow well in clay soils.

What are fossils?

Scientists use fossils to learn about the past. A **fossil** is evidence of an organism that lived long ago. Most fossils are found in sedimentary rock. Sediment can bury plant or animal remains. As the sediment turns to rock, the remains become fossils.

Shells often leave behind fossils known as molds. A mold is a hollow form with a certain shape. A cast is a fossil that is formed or shaped in a mold fossil.

There are many other kinds of fossils. An imprint is a fossil made by pressing. A footprint is an example. Wood or bone can become petrified. This means it has turned to stone.

The remains of an entire organism can become a fossil. Large animals called mammoths have been found. Their bodies were found frozen in ice. Insects have been found stuck in hardened tree sap.

Kinds of Fossils

foot imprint

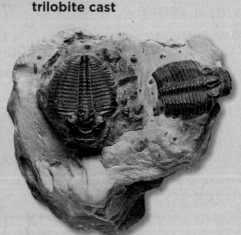
trilobite cast

petrified wood

✓ *Quick Check*

14. A(n) _____ fossil is made by pressing.

How do we study fossils?

Scientists compare fossils to plants and animals living today. The fossils show how living things change over time.

Earth's land and climate also change over time. Most changes are very, very slow. Scientists measure Earth's history in millions—even billions—of years. The time used to measure Earth's history is called geologic time. Fossils help scientists study geologic time.

Evidence of Change

Fossils are found in rock layers. Younger fossils are in a top layer. Older fossils are in a lower layer. The fossils show how that part of Earth changed over time.

Sometimes scientists find a fish fossil on land. Then they know that the land was once covered by water. Scientists may also find plant fossils in cold places where no plants grow today. The fossils show that the climate there was once warmer.

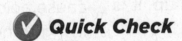 **Quick Check**

15. What is geologic time?

_____

_____

◄ Scientists must work carefully to clean and prepare fossils for study.

What are fossil fuels?

A **fossil fuel** is an energy source that formed millions of years ago. Fossil fuels formed from the remains of buried plants and animals. Coal, oil, and natural gas are fossil fuels. Most of the energy used to make electricity comes from fossil fuels. Most of the energy to drive vehicles also comes from fossil fuels.

Fossil fuels are nonrenewable resources (non•ri•NEW•i•buhl REE•sor•sez). A **nonrenewable resource** is a useful material that cannot be replaced easily. Once it is used up, it is gone forever.

Finding Fossil Fuels

Finding fossil fuels can be hard. They are deep under Earth's surface. People dig to find fossil fuels. It is expensive to find and produce fossil fuels.

Using Fossil Fuels

One liter of natural gas takes millions of years to form. It burns within seconds. Burning fossil fuels releases energy. Burning fossil fuels also causes air pollution.

People who breathe the polluted air may become sick. Polluted air can cause acid rain. Acid rain can harm living things.

How Coal Forms

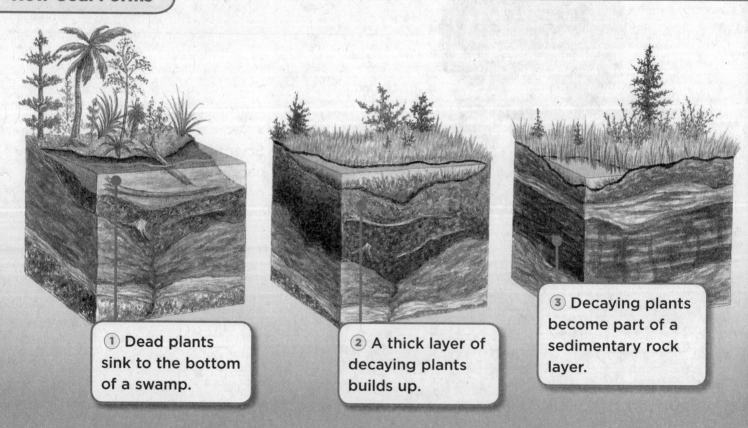

① Dead plants sink to the bottom of a swamp.

② A thick layer of decaying plants builds up.

③ Decaying plants become part of a sedimentary rock layer.

Alternative Energy

No one knows how long our fossil fuel supply will last. So scientists look for other ways to get energy. These other ways are called alternative energy sources.

 Quick Check

16. What is a fossil fuel?

_____

_____

17. What is a nonrenewable resource?

_____

_____

▲ This machine is an oil pump. It gets oil from below the ground.

Read a Diagram

What happens after the decaying plants become part of a sedimentary rock layer?

_____

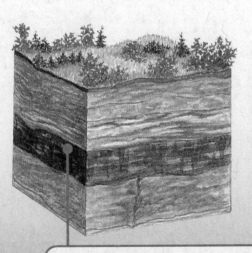

④ The rock layer is pressed into soft coal, a fossil fuel.

⑤ Under intense heat and pressure, the soft coal turns to hard coal. It is also a fossil fuel.

What can we use instead of fossil fuels?

Earth gives us many renewable resources. A **renewable resource** is a useful material that is replaced in nature.

The Sun gives us energy every day. A solar cell is a tool that collects energy from sunlight.

Windmills use energy from wind to make electricity. We can also get energy from:

• flowing water
• ocean tides
• the heat below Earth's surface

Wind farms gather the energy from wind. They turn it into electricity. ▼

Energy Sources in the United States

Look at the pie chart. It shows where the United States got energy for electricity in 2005. Just 9 percent came from renewable resources. The rest came from nonrenewable resources.

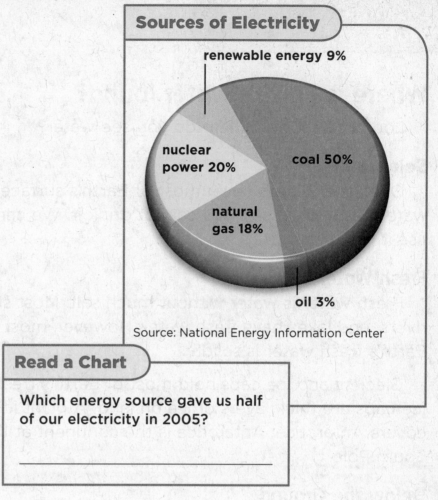

Sources of Electricity

renewable energy 9%

nuclear power 20%

coal 50%

natural gas 18%

oil 3%

Source: National Energy Information Center

Read a Chart

Which energy source gave us half of our electricity in 2005?

_____

Quick Check

Tell whether the sentences are true or false. If false, correct the sentence.

18. A renewable resource is gone forever once it is used.

_____

19. A solar cell collects energy from sunlight.

_____

20. We can get energy from the wind using windmills.

_____

Where is Earth's water found?

Look at a globe. Where do you see water?

Salt Water

Oceans and seas cover most of Earth's surface. Ocean water has a lot of salt. We cannot drink it. We cannot use it for farming.

Fresh Water

Fresh water is water without much salt. Most streams, rivers, and lakes have fresh water. However, most of Earth's fresh water is solid!

Glaciers and ice caps hold most of Earth's fresh water. Ice caps are thick layers of ice on land. A giant ice cap covers Antarctica. Antarctica is the continent at the South Pole.

Below the Ground

When it rains, water soaks into the soil. Plants use some of this water. The rest of the water travels down to rocks below the soil. The water becomes groundwater. Groundwater is the water stored in cracks and spaces under the ground.

Most of Earth's fresh water is solid ice.

Watersheds

On land, water may flow downhill into a common stream, lake, or river. These areas are called watersheds. People who live in a watershed use the water that drains through it.

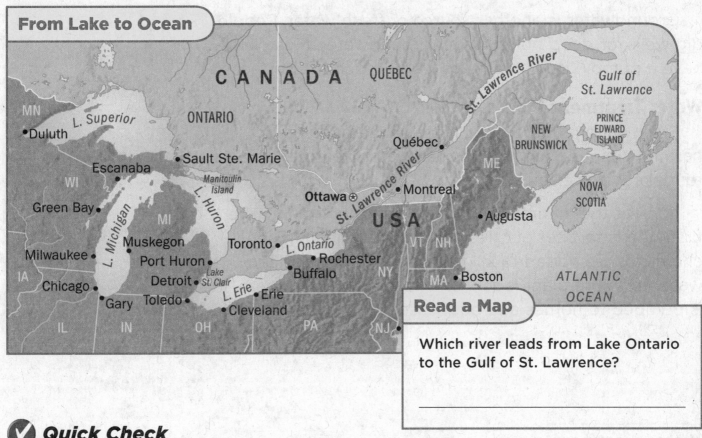

From Lake to Ocean

Read a Map

Which river leads from Lake Ontario to the Gulf of St. Lawrence?

_____

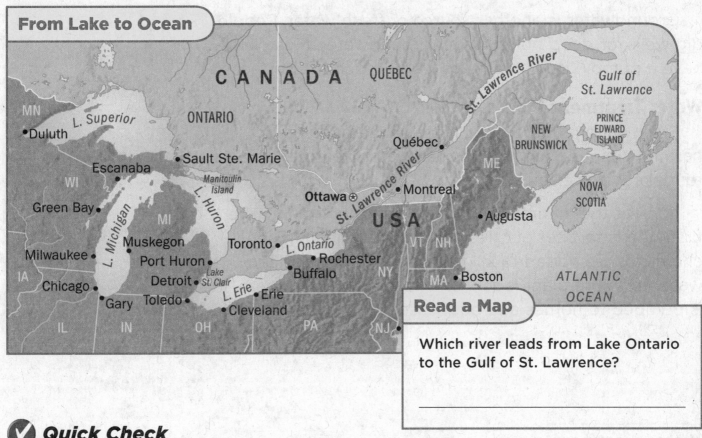

✓ Quick Check

Tell if the sentences are true or false. If false, correct the sentence.

21. Ocean water has no salt.

_____

22. Most fresh water is frozen, or solid.

_____

23. Groundwater is stored in lakes.

_____

How is fresh water supplied?

Most large towns and cities get their water from reservoirs (REZ•uhr•vwahrz). A **reservoir** is a storage area for holding fresh water. Some reservoirs are natural lakes or ponds. Others are built by people. Pipelines supply people with water from reservoirs.

Groundwater is another source of fresh water. People dig wells, or deep holes, to get groundwater. Most wells have a pump. The pump helps water reach the surface.

Water Treatment Plants

Fresh water can be dirty. It may have bacteria or harmful chemicals in it. Cities use water treatment plants to make water clean and pure.

Look at the diagram of a water treatment plant. The first step is to filter out trash and large objects. Then chemicals are added to kill harmful organisms. Clean water from the plant is sent to a reservoir. From there, it is pumped to homes and businesses.

Water Treatment

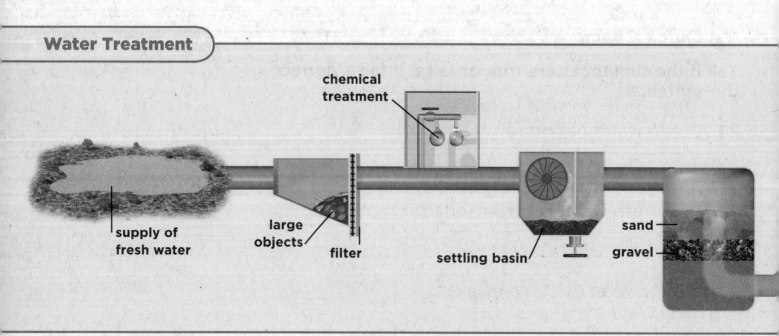

chemical treatment

supply of fresh water

large objects

filter

settling basin

sand

gravel

Saving Earth's Resources

Lake Mead is a reservoir shared
by Arizona and Nevada.

✔ Quick Check

24. Name two sources of fresh water.

_____, _____

25. Why are chemicals added to water in a treatment
plant?

_____

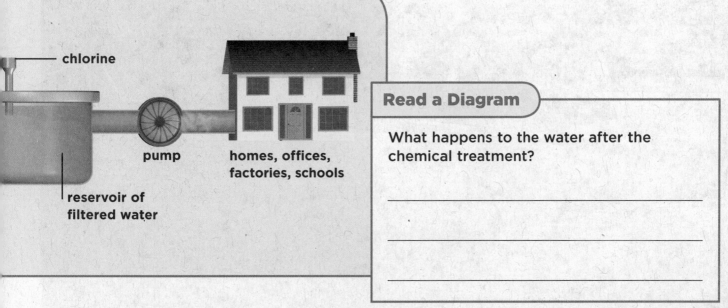

chlorine

pump

homes, offices,
factories, schools

reservoir of
filtered water

Read a Diagram

What happens to the water after the
chemical treatment?

_____

_____

_____

How do we use water?

People use Earth's water in all sorts of ways. Fresh water is used in farming. In some places, irrigation (eer•uh•GAY•shuhn) supplies the water for growing crops. Irrigation is a way to bring water into the soil through pipes or ditches.

Water is also used for fun. People canoe and raft on fast-moving rivers. On lakes people swim and waterski. Fishing is done in fresh water and salt water. In winter many people skate on frozen lakes.

▲ Some farms grow plants in water instead of soil!

▼ People use water for fun and recreation.

Water is important to industry. It is used to make many products, such as paint and paper. Water is used to make electricity. Ships use water to move goods all over the world.

✔ Quick Check

Name three ways people use water.

26. _____

27. _____

28. _____

▼ Waterways help people move goods from place to place.

LOG ON **ⓔ-Review** Summaries and quizzes online at **www.macmillanmh.com**

What is pollution?

The things that make up an area—air, water, and land—form an environment. All living things need a healthy environment. A healthy environment has clean air, water, and land.

When a harmful substance is put into an environment, it causes **pollution** (puh•LEW•shuhn). Most pollution is made by people.

Air Pollution

When we burn fossil fuels, gases go into the air. Some of the gases mix with water droplets in the air. This can make acid rain. Acid rain can hurt living things and buildings.

Other gases can form smog. Smog hangs in the air like a cloud. Smog makes it hard to breathe.

Cars, trucks, and buses are major sources of air pollution in cities. ▼

Water and Land Pollution

People cause water pollution when they dump trash into oceans, lakes, and rivers. Water can also be polluted by chemicals. Chemicals used on gardens can wash into rivers. Oil spills from ships can pollute the ocean. Polluted water can kill plants and animals.

Like water, land can be polluted by chemicals. Land is also polluted when people litter, or throw trash on the ground. Littering is against the law.

▲ People cause land pollution by leaving tires, paper, and other trash on the ground.

 **Quick Check**

29. What is pollution?

_____

30. How does smog hurt people?

_____

31. Give one reason why we should never litter.

_____

How can we protect the soil and water?

Conservation (kon•sur•VAY•shuhn) means using resources wisely.

Farmers use methods to slow erosion. They keep the soil in place to support plants. On sloping lands, farmers plow fields in curved rows that follow the shape of the land. This is called contour plowing.

People can conserve soil by using compost in their gardens. Compost is made from dead plant and animal matter. Food scraps, fallen leaves, and cut grass can be used to make compost.

People can conserve water in many ways. Cities can collect used water. The used water can be cleaned and released. You can conserve water by turning off a faucet when you are not using it.

Contour Plowing

Read a Photo

How can you tell that this area has sloping land?

_____

_____

✔ *Quick Check*

32. Using resources wisely is called

_____ .

The 3 Rs in Action

What are the 3 Rs?

Three ways to conserve resources start with the letter R.

Reduce

To **reduce** means to use less of something. This is the simplest way to conserve.

Reuse

To **reuse** means to use something over again. For example, you can reuse old clothes as cleaning rags.

Recycle

To **recycle** means to make a new thing from old materials. Recycling keeps things in use and out of landfills.

Read a Photo

Which way to conserve resources does the last picture show?

_____

✓ *Quick Check*

Name the 3 Rs of conservation. Give an example of each.

33. _____

_____

34. _____

_____

35. _____

_____

Saving Earth's Resources

Choose the letter of the best answer.

1. A natural, nonliving material that makes up rock is called a(n)
   a. mineral.
   b. fossil.
   c. igneous rock.
   d. pollution.

2. Rock formed from sediments that are stuck together is a(n)
   a. fossil.
   b. igneous rock.
   c. metamorphic rock.
   d. sedimentary rock.

3. Rock that is formed when melted rock cools is
   a. metamorphic rock.
   b. sedimentary rock.
   c. a mineral.
   d. igneous rock.

4. A storage area for holding fresh water is a
   a. nonrenewable resource.
   b. renewable resource.
   c. reservoir.
   d. mineral.

5. Harmful or unwanted material that has been added to the environment is called
   a. humus.
   b. metamorphic rock.
   c. pollution.
   d. conservation.

6. Rock that is formed from another kind of rock through heat and pressure is
   a. metamorphic rock.
   b. a fossil.
   c. igneous rock.
   d. sedimentary rock.

7. The remains of an organism that lived long ago is called a
   a. reservoir.
   b. fossil.
   c. sedimentary rock.
   d. mineral.

8. Using resources wisely is known as
   a. pollution.
   b. conservation.
   c. the rock cycle.
   d. permeability.

Match the words in the first column to the best answer in the second column.

1. rock cycle _d_

2. recycle _g_

3. nonrenewable _a_

4. pore space _e_

5. humus _k_

6. bedrock _h_

7. renewable resource _i_

8. reuse _f_

9. permeability _b_

10. reduce _c_

11. fossil fuel _j_

a. cannot be replaced easily

b. tells how fast water goes through a porous material

c. to use less of something

d. the process in which rocks change from one type to another

e. the space between particles of soil

f. to use something over again

g. to make a new product from old materials

h. the solid layer of rock below the soil

i. a useful material that is replaced quickly in nature

j. an energy resource that formed millions of years ago

k. dead plant or animal matter that has broken down

Answer the following question. Use words from the first column for the answers.

12. What are the 3 Rs? _Recycle_   _Reuse_   _Reduce_

Summarize

_____

_____

_____

_____

CHAPTER 7

Weather and Climate

Vocabulary

 **atmosphere** the blanket of gases that surrounds Earth

 **temperature** how hot or cold something is

 **humidity** a measure of how much water vapor is in the air

 **air pressure** the force of air pushing on an area

 **barometer** a tool that measures air pressure

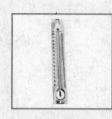 **thermometer** a tool that measures temperature

 **evaporation** a liquid changing to a gas

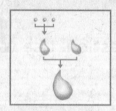 **water vapor** water in the gas state

 **condensation** the process of a gas changing to a liquid

 **cloud** a collection of tiny water droplets or ice crystals that hangs in the air

What are weather and climate?

 precipitation water that falls from clouds down to Earth

 **water cycle** the movement of water between Earth's surface and the air

 **air mass** a large body of air that has similar properties throughout

 **front** a boundary between air masses with different temperatures

 **warm front** a warm air mass pushing into a cold air mass

 **cold front** a cold air mass pushing under a warm air mass

 **stationary front** a boundary between air masses that are not moving

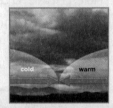 **forecast** to predict the weather

 **climate** the average weather pattern of a region over time

 **current** a flow of a gas or a liquid

What is in the air?

Air surrounds Earth like a thin blanket. This blanket of air is the **atmosphere** (AT•muhs•feer).

Gases

The atmosphere is a mix of different gases. Most of the atmosphere is made of nitrogen (NYE•truh•juhn) and oxygen. Without these gases, plants and animals could not live on Earth.

The atmosphere also has other gases. Carbon dioxide and water vapor are two other gases in the atmosphere.

You cannot see the gases that make up the atmosphere. However, you can feel them when the wind blows.

Layers of Earth's Atmosphere

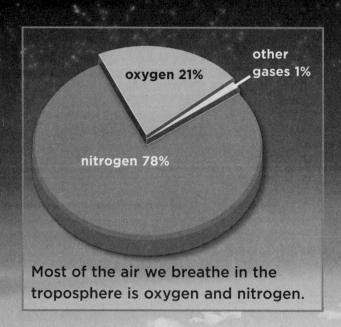

oxygen 21%

other gases 1%

nitrogen 78%

Most of the air we breathe in the troposphere is oxygen and nitrogen.

Layers of the Atmosphere

Earth's atmosphere is made up of layers.

- The troposphere (TROHP•uh•sfeer) is the layer closest to Earth's surface. This is where weather takes place.
- The stratosphere (STRAT•uh•sfeer) is above the troposphere. In this layer, gas particles get farther apart.
- The mesosphere (MEZ•uh•sfeer) is next. Temperatures here get colder as you go up.
- The thermosphere (THURM•uh•sfeer) is the top layer. The thermosphere has fewer particles of gas than lower layers. The particles are farther apart from one another.

 Quick Check

1. Which gas makes up most of the atmosphere?

_____

650+ km

thermosphere

85 km

mesosphere

50 km

stratosphere

17 km

troposphere

Read a Diagram

Which layer of the atmosphere is closest to Earth's surface?

_____

What are some properties of weather?

Weather is the condition of the atmosphere at a given time and place.

Air Temperature and Humidity

Temperature (TEM•puhr•uh•chuhr) describes how hot or cold something is. When the air temperature changes, air moves. Moving air is called wind.

If the air feels damp and sticky, it is humid. **Humidity** (hyew•MID•i•tee) is a measure of how much water vapor is in the air. Deserts have low humidity. Rain forests have high humidity.

Humidity in a Rain Forest

Read a Photo

What in the picture makes you think that the humidity in the rain forest is high?

_____

_____

Mountain climbers use special equipment to deal with low temperature and low air pressure.

Air Pressure and Precipitation

The force of air pushing down on an area is called **air pressure.** Temperature affects air pressure. Particles of cold air are close together. This makes the pressure higher. Altitude also affects air pressure. There are fewer air particles in the air at higher altitudes than at lower altitudes. Air pressure will be higher at the ocean shore than on a mountain top.

Any liquid or solid water that falls from clouds is precipitation (pree•sip•uh•TAY•shuhn). Types of precipitation include rain, snow, sleet, and hail.

 **Quick Check**

Name four properties of weather.

2. _____

3. _____

4. _____

5. _____

How can you measure weather?

Weather scientists get information at a weather station. You can set up your own weather station with these tools.

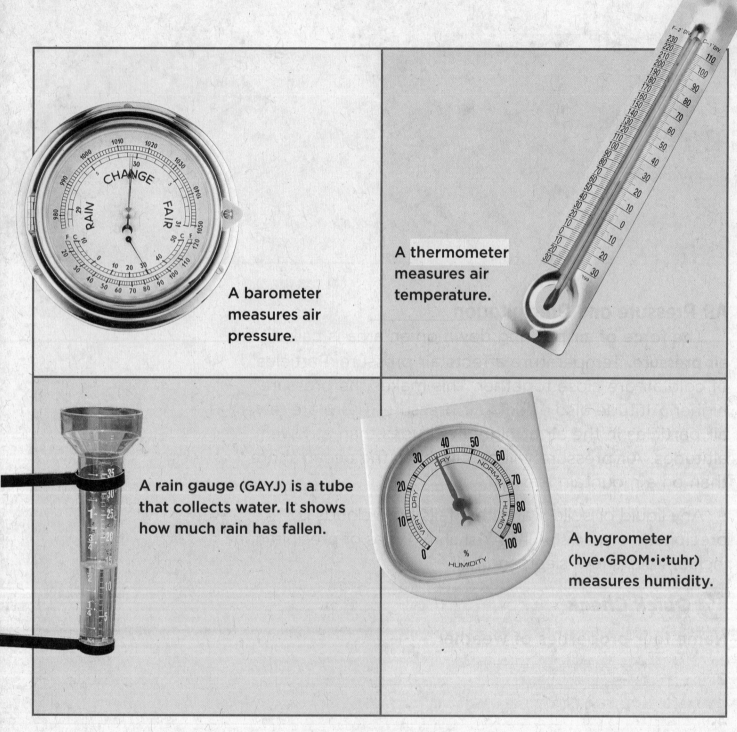

A barometer measures air pressure.

A thermometer measures air temperature.

A rain gauge (GAYJ) is a tube that collects water. It shows how much rain has fallen.

A hygrometer (hye•GROM•i•tuhr) measures humidity.

A wind vane points in the direction from which the wind is blowing.

An anemometer (an•uh•MOM•i•tuhr) measures wind speed. The faster the wind blows, the faster the cup spins.

✓ Quick Check

Match the weather tools to what they measure.

6. _e_ thermometer   **a.** measures humidity

7. _f_ wind vane   **b.** measures amount of rainfall

8. _b_ rain gauge   **c.** measures wind speed

9. _a_ hygrometer   **d.** measures air pressure

10. _c_ anemometer   **e.** measures air temperature

11. _d_ barometer   **f.** shows direction of the wind

Why does water change state?

Water moves from Earth's surface to the air. Then it moves back to the surface. Water changes state as it moves. A change in temperature can cause a change in state.

Evaporation

Evaporation happens when water changes from a liquid to a gas.

Water evaporates (ee•VAP•uh•rayts) from oceans, streams, lakes, rivers, and ponds. The Sun's heat causes particles of liquid water to warm up. The particles move faster. Some of the particles rise into the air as water vapor. **Water vapor** is water in the gas state.

How Water Changes State

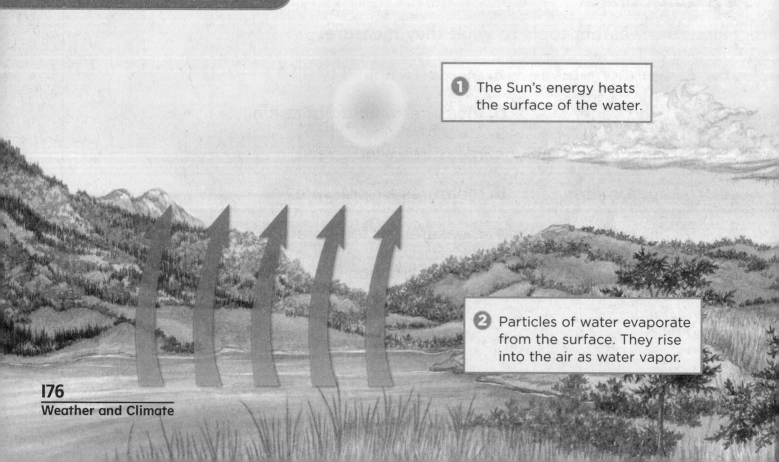

① The Sun's energy heats the surface of the water.

② Particles of water evaporate from the surface. They rise into the air as water vapor.

Condensation

The process of a gas changing to a liquid is called **condensation.** When water vapor cools, it condenses (kuhn•DEN•sez) to liquid water. In the air, water condenses on dust particles. This forms a cloud. A **cloud** is a group of water droplets in the atmosphere.

Precipitation

Precipitation (pri•sip•i•TAY•shuhn) is solid or liquid water that falls from clouds down to Earth. Inside a cloud, water droplets may grow larger. Water droplets may also freeze into ice. To freeze is to change from a liquid to a solid. Water droplets and bits of ice can become too heavy. They fall to Earth's surface as precipitation.

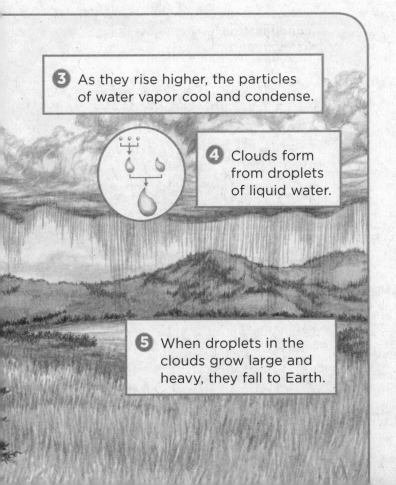

3 As they rise higher, the particles of water vapor cool and condense.

4 Clouds form from droplets of liquid water.

5 When droplets in the clouds grow large and heavy, they fall to Earth.

✅ Quick Check

12. Water changing from a liquid to a gas is _____.

13. A gas changing to a liquid is _____.

14. Water falling from clouds down to Earth is _____.

Where does water go?

Water is always moving. The **water cycle** is the movement of water between Earth's surface and the air. Evaporation, condensation, and precipitation help water move through the water cycle.

In the Air

The Sun powers the water cycle. Sunlight heats water on Earth's surface. The water evaporates. Water vapor rises into the air and cools. Then water condenses. Clouds form. Precipitation falls from the clouds.

The Water Cycle

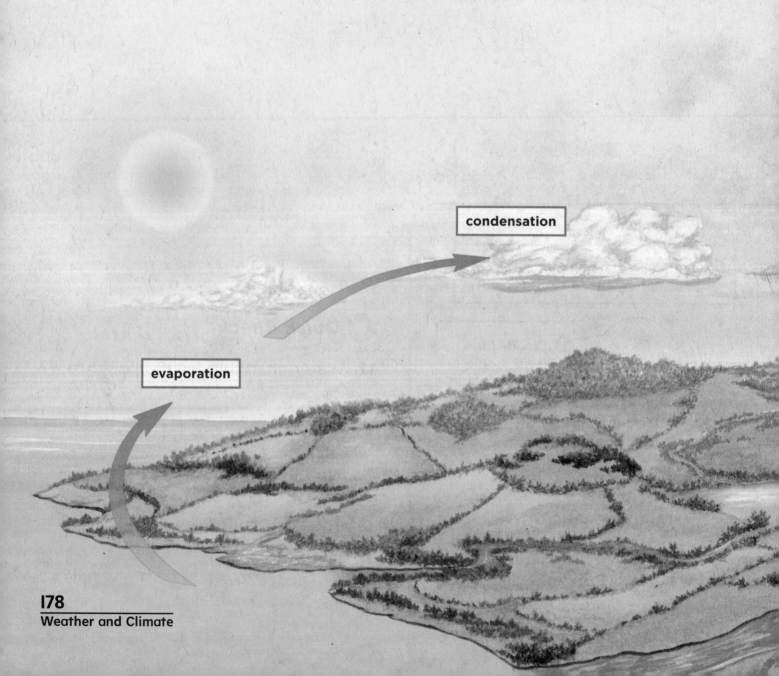

condensation

evaporation

On and Below the Ground

Precipitation can collect in oceans, lakes, rivers, glaciers, and ice caps. Rain flows over Earth's surface as runoff. Runoff goes into lakes, oceans, and rivers. Rain can also soak into the ground. Some of this water is used by plants and some evaporates. The rest becomes groundwater.

LOG ON *Science in Motion* Watch how the water cycle works at www.macmillanmh.com

 **Quick Check**

15. The movement of water between Earth's surface and the air is the

_____.

Read a Diagram

Which step happens after water evaporates?

_____

precipitation

runoff

transpiration

evaporation

groundwater

What are some types of clouds?

Clouds form at different heights above Earth's surface. Clouds are classified based on how and where they form. There are three main types of clouds. Sometimes more than one type of cloud can be seen in the sky.

Cumulus

Cumulus (KYEW•myuh•luhs) clouds are puffy, white clouds. They look like cotton balls.

- They often have a flat bottom.
- Cumulus clouds may grow dark and thick.
- Dark and thick cumulus clouds are called cumulonimbus (kyew•myuh•loh•NIM•buhs) clouds.
- Cumulonimbus clouds cause precipitation.

Cloud Types

cumulus

stratus

cirrus

Stratus

Stratus (STRAT•uhs) clouds form in layers. They look like sheets or blankets.

- They are often the lowest clouds in the sky.
- Fog is a stratus cloud on or near Earth's surface.
- Stratus clouds can cause precipitation.

Cirrus

Cirrus (SIR•uhs) clouds look thin and feathery.

- They are made of tiny bits of ice.
- They are usually found very high in the sky.
- Cirrus clouds often appear before a storm.

✔ Quick Check

Circle the correct answer.

16. Which cloud type looks thin and feathery?

　　a. cirrus

　　b. cumulus

　　c. stratus

Many Types of Clouds

cirrus

cirrocumulus

altocumulus

cumulonimbus

Read a Diagram

Which type of cloud has some properties of cirrus clouds and some properties of cumulus clouds?

_____

What are other forms of precipitation?

Rain is the most common form of precipitation. Snow, sleet, and hail are other forms.

Remember, to freeze is to change from a liquid to a solid. To melt is to change from a solid to a liquid. Bits of ice can collect in a cloud. If the bits get heavy, they fall as snow.

Snow can melt as it falls from a cloud. Warm air or sunlight can heat the snow. The heat can change the snow to rain.

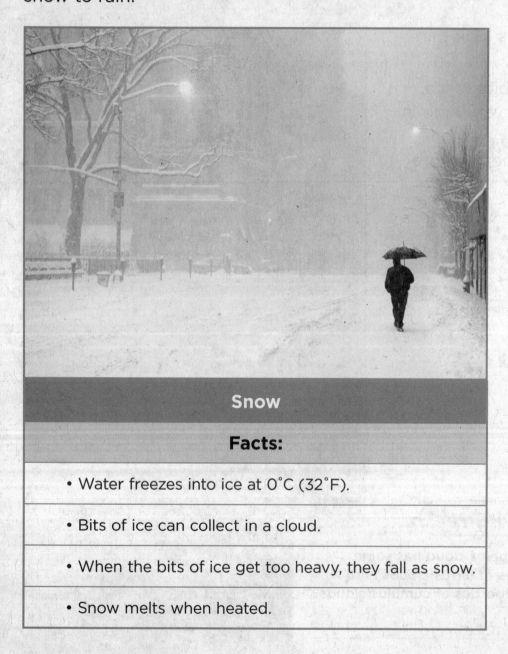

Snow

Facts:

• Water freezes into ice at 0°C (32°F).
• Bits of ice can collect in a cloud.
• When the bits of ice get too heavy, they fall as snow.
• Snow melts when heated.

Sleet and Hail

Facts:

- Liquid rain can freeze and become sleet or hail.

- Rain that turns into small chunks of ice as it falls is called sleet.

- Hail is ice, but it is much larger than sleet.

- Hail forms inside the tall clouds of a thunderstorm.

▼ Large hail is very dangerous.

Sometimes rain freezes as it falls. The rain turns to ice. This ice is called sleet. Hail is made of ice, too. The ice chunks are larger than sleet. Hail forms inside the clouds of a thunderstorm.

✔ Quick Check

17. Does water freeze or melt to form snow, sleet,

and hail? _____

What are air masses and fronts?

Weather is different in different places. Some places get lots of rain. Some places are dry, like deserts. In other places sunny days follow rainy days. Why does this happen?

Weather changes when air masses move. **Air masses** are large bodies of air. Air masses meet along fronts.

Air Masses

The temperature of the air in an air mass is similar throughout. The humidity is also similar. Weather is the same in all parts of an air mass.

Air masses form all the time. They usually form near the equator or poles. They move across Earth. The map shows some paths they often take.

Air Masses in North America

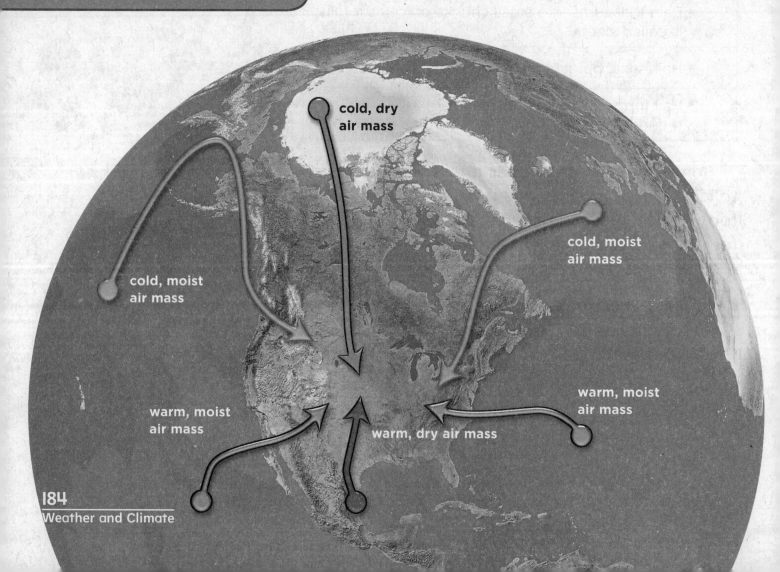

cold, dry air mass

cold, moist air mass

cold, moist air mass

warm, moist air mass

warm, dry air mass

warm, moist air mass

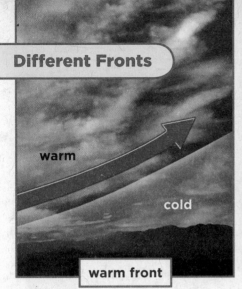

warm

cold

warm front

warm

cold

cold front

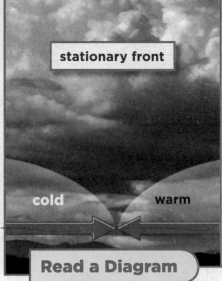

stationary front

cold   warm

Different Fronts

Fronts

The area where two air masses meet is called a front. A **front** is the boundary between two air masses that have different temperatures. Fronts usually cause a change in the weather.

At a **warm front,** a warm air mass pushes into a cold air mass. The warm air slides up and over the cold air. This kind of front often brings light, steady rain. After the front passes, the air temperature rises.

At a **cold front,** a cold air mass pushes under a warm air mass. The cold air forces the warm air upward quickly. Cold fronts often bring stormy weather.

A **stationary front** is a boundary between air masses that are not moving. This kind of front can cause rainy weather for days.

 Quick Check

Write the name of each front on the line.

Weather		Front
Warm air mass pushes into a cold air mass.	→	18. _____
Cold air mass pushes under a warm air mass.	→	19. _____
Air masses do not move.	→	20. _____

Read a Diagram

Where does the warm air mass move in a cold front?

_____

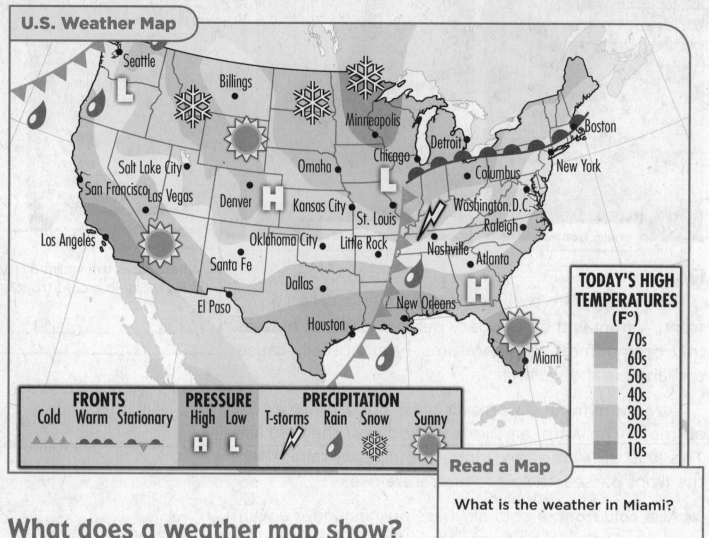

U.S. Weather Map

Seattle
Billings
Minneapolis
Detroit
Boston
Chicago
Columbus
New York
Salt Lake City
Omaha
San Francisco
Las Vegas
Denver
Kansas City
St. Louis
Washington.D.C.
Raleigh
Los Angeles
Oklahoma City
Little Rock
Nashville
Santa Fe
Atlanta
Dallas
New Orleans
El Paso
Houston
Miami

TODAY'S HIGH TEMPERATURES (F°)

	70s
	60s
	50s
	40s
	30s
	20s
	10s

FRONTS
Cold Warm Stationary

PRESSURE
High Low
H L

PRECIPITATION
T-storms Rain Snow   Sunny

Read a Map

What is the weather in Miami?

_____

What does a weather map show?

Weather maps give information about the weather. Weather maps show:

- air temperature
- high and low air pressure
- precipitation
- wind
- the locations of fronts

The locations of thunderstorms are shown on weather maps. ▶

186
Weather and Climate

Forecasting

Scientists use weather maps to make forecasts. To **forecast** is to predict the weather. Temperature, air pressure, and the direction of moving fronts give clues to future weather.

Look at the weather map. One forecast based on this map might be a chance of thunderstorms in Nashville.

Weather maps can forecast snow storms.

 Quick Check

21. What information might you find on a weather map?

_____

_____

_____

_____

What are the signs of severe weather?

There are many types of storms. They include:

- **Thunderstorms:** Thunder is a loud sound made when lightning heats the air around it quickly. Thunder tells you that a thunderstorm is near.
- **Tornadoes:** A tornado is a rotating column of air that touches the ground. You need to take cover if you see one.
- **Hurricanes:** Hurricanes form over warm ocean water. They bring very heavy rains and strong winds.

▲ Strong winds and lightning can make a storm dangerous.

Storm Safety

There are things you can do to stay safe in severe weather.

- **Thunderstorms:** Stay away from water and trees.
- **Tornadoes:** Head for a strong shelter, such as a basement.
- **Hurricanes:** Move inland.
- In any storm, always listen for directions. Be sure to follow warnings on the radio and television.

▲ During a hurricane, move inland.

◀ During a tornado, head for a strong shelter.

✔ Quick Check

What should you do to stay safe in each kind of storm?

22. thunderstorm _____

_____

23. tornado _____

_____

24. hurricane _____

_____

What is climate?

The average weather pattern of a region over time is called **climate** (KLYE•mit). Climate is not the same everywhere. The climate in Arizona is warm and dry all year. Snow and rain rarely fall. The climate in Alaska is cool. Snow is common there.

Canada

temperate

Antarctica

polar

Arizona

dry

Climate Regions

A region of land can have certain patterns of temperature, humidity, precipitation, and wind. Scientists call such a region a climate region.

Polar regions have cold climates with little precipitation. Temperate regions can have four seasons. However, some may have just a dry season and a rainy one. Tropical regions are warm and rainy.

Ecuador

tropical

Alaska

cool

✔ Quick Check

25. What is climate?

_____

_____

What determines climate?

Several things affect a climate region over time.

Latitude

Thin lines that run east and west across some maps are lines of latitude. Latitude is a measure of how far a place is from the equator (0° latitude).

Climates near the equator are often warm and rainy. Between the equator and the poles, the climate is mild. Near the poles, the climate is cold all year.

Global Winds

Temperature differences between latitudes cause global winds. These winds move air between the equator and the poles.

Warm air near the equator rises and moves toward the poles. Cold air near the poles sinks and moves toward the equator.

Ocean Currents

A **current** is a flow of a gas or a liquid. Some ocean currents move warm water from the equator to the poles. Others move cold water from the poles toward the equator.

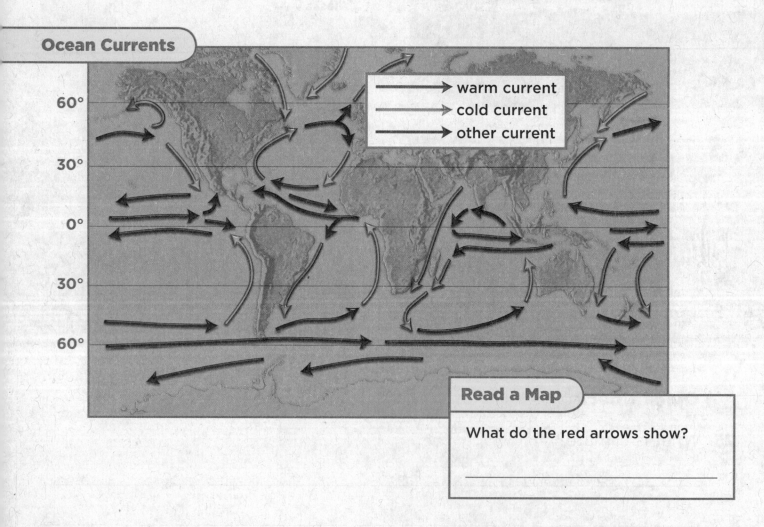

Ocean Currents

→ warm current
→ cold current
→ other current

60°
30°
0°
30°
60°

Read a Map

What do the red arrows show?

_____

Distance from Water

Land and water warm up and cool down at different speeds. Water gets hotter more slowly than land does. It also cools more slowly. These differences affect the air temperature and precipitation near water.

Climates near lakes and oceans are different than climates farther inland. Climates near water:

• have more clouds and rain
• have cooler summers
• have warmer winters

✓ Quick Check

Write true or false for each sentence. If it is false, correct the sentence.

26. A current is a flow of a gas or a liquid.

_____

27. Climates near the equator are cold and dry.

_____

28. Water heats up more quickly than land.

_____

▼ Indiana is an inland state. Winters there are cold and snowy.

How do mountains affect climate?

Latitude, water, and wind are not the only things that affect climate. Mountains also affect climate.

Altitude

Altitude is how high a place is above sea level. Air is colder at higher altitudes.

Altitude changes as you move up a mountain. Climate at the bottom of a mountain is always warmer than at the top.

The Mountain Effect

Read a Photo

Where is the air coolest on this mountain?

_____

Clouds and Precipitation

When an air mass meets a mountain, it moves up the side. As the altitude gets higher, the temperature gets cooler. Water vapor in the air condenses. Clouds form. The water droplets in the clouds get heavy. Precipitation falls.

By the time the air mass gets over the mountain, the air is dry. This causes one side of a mountain to have a wet climate. The other side often has a dry climate.

An air mass loses moisture as it moves over a mountain.

◄ Clouds form as air cools when it moves over the mountain.

✓ Quick Check

29. What happens as an air mass moves up a mountain?

_____

_____

_____

Weather and Climate

Use a word from the box to name each example described below. Then find the same words in the puzzle.

air pressure
atmosphere
cold front
evaporation
temperature
thermometer
water vapor

1. How hot or cold something is _____

2. The force of air pushing on an area _____

3. A tool that measures temperature _____

4. The process of liquid changing to a gas _____

5. Water in the gas state _____

6. The blanket of gases that surrounds Earth _____

7. A cold air mass pushing under a warm air mass _____

```
C A L A F L I H F W S A P
O T T L N M O F E A H I R
L M T E M P E R A T U R E
D O P H D V F O A E S P R
F S K B C U I N H R A R E
R P Y R Z L C T U V O E A
O H Y R D M O G M A S S J
N E R T D V U U I P O S O
T R Y Z N J W G D O H U P
A E R R S W H J I R O R I
T H E R M O M E T E R E D
P O F T F R C E Y M U F N
D E V A P O R A T I O N E
```

Match each word with its meaning. Write the letter in the blank.

1. ___ condensation
2. ___ barometer
3. ___ climate
4. ___ current
5. ___ water cycle
6. ___ precipitation
7. ___ air mass
8. ___ stationary front
9. ___ forecast
10. ___ warm front

a. water that falls from clouds down to Earth

b. the average weather pattern of a region over time

c. a large body of air that has similar properties throughout

d. a tool that measures air pressure

e. a warm air mass pushing into a cold air mass

f. a boundary between air masses that are not moving

g. the movement of water between Earth's surface and the air

h. to predict the weather

i. a flow of a gas or a liquid

j. the process of a gas changing to a liquid

Summarize

_____

_____

_____

_____

_____

CHAPTER 8

The Solar System and Beyond

Vocabulary

 **rotation** the spinning of an object around a center point

 **axis** a real or imaginary line that an object spins around

 **revolution** one complete trip around an object in a circular or nearly circular path

 **crater** a hollow area or pit in the ground

 **phase** an apparent change in the Moon's shape

 **solar eclipse** the Moon making a shadow on Earth

 **lunar eclipse** Earth making a shadow on the Moon

What objects are in the solar system and beyond?

solar system the Sun and all the objects that travel around it

planet a round object in space that travels around the Sun

gravity a force of attraction between all objects

telescope a tool that makes objects look closer and larger

comet a chunk of ice, rock, and dust that moves around the Sun

star a ball of hot gases that gives off light and heat

constellation a group of stars that make a pattern in the night sky

What causes day and night?

Earth is always moving. It spins, or rotates. **Rotation** (roh•TAY•shuhn) is the act of spinning in a circle. Earth rotates around a center point. This center point is Earth's axis (AK•sis). An **axis** is a real or imaginary line that an object spins around. The dotted line in the picture is Earth's axis.

Earth's rotation causes day and night. Half of Earth always faces the Sun. It is day there. The other half is dark. It is night there. As Earth spins, the part facing the Sun moves. Soon the day will end. Night for that part of Earth will begin.

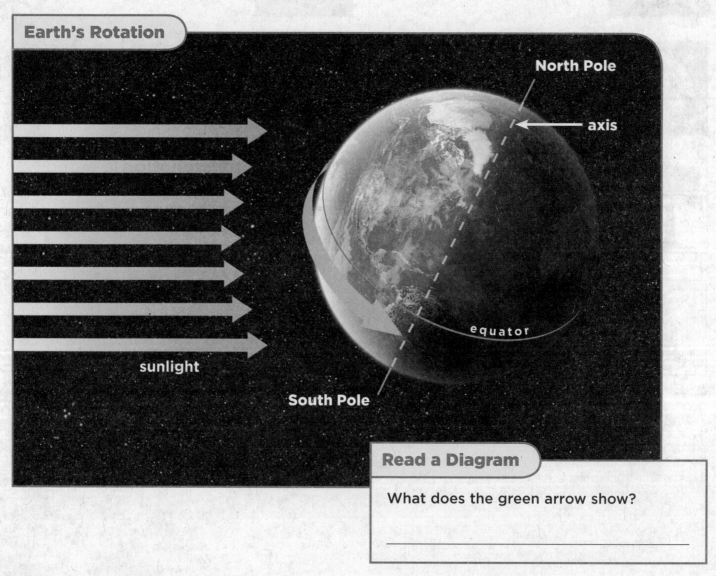

Earth's Rotation

North Pole

axis

equator

sunlight

South Pole

Read a Diagram

What does the green arrow show?

_____

Apparent Motion

Earth moves around the Sun. However, during the day it looks like the Sun moves around Earth. This is the Sun's apparent motion. Apparent motion is the way something seems to move. Apparent motion is not real motion.

Earth's rotation causes the Sun's apparent motion. Earth's rotation also makes it look like the stars move. Instead, it is Earth that moves. As Earth spins, it looks like objects in space are moving across the sky.

Shadows

A shadow forms when light is blocked. The light hits an object but cannot pass through. Blocked sunlight makes shadows on Earth. As the Sun moves, shadows change during the day.

▲ When the Sun is high in the sky, this antelope has a shorter shadow.

▲ When the Sun is low in the sky, the antelope has a longer shadow.

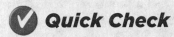 **Quick Check**

Fill in the missing cause and effects in the chart.

Cause ⟶ Effect	
A part of Earth turns away from the Sun.	⟶ 1. _____
2. _____	⟶ It becomes day.
Light is blocked.	⟶ 3. _____

What causes seasons?

Earth revolves (ri•VAHLVZ) around the Sun. **Revolution** is the movement of one object around another. The path of a revolving object is its orbit. Earth's orbit around the Sun takes $365\frac{1}{4}$ days. This is one year.

Earth's axis is titled at an angle. The tilt points in the same direction all year. Sunlight hits Earth at different angles because of this tilt. The seasons are caused by both Earth's tilted axis and its revolution.

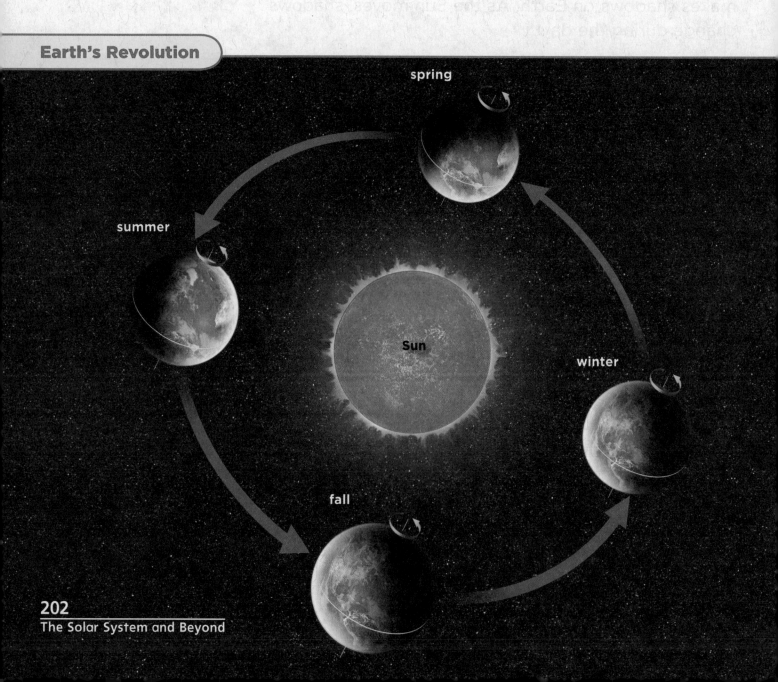

Earth's Revolution

spring

summer

Sun

winter

fall

Seasons in the Northern Hemisphere

In June the North Pole tilts toward the Sun. Strong sunlight hits the Northern Hemisphere. It is summer there.

In December the North Pole tilts away from the Sun. Weak sunlight hits the Northern Hemisphere. It is winter there.

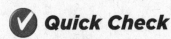 **Quick Check**

4. When Earth travels around the Sun, it is called a

_____.

5. The path a revolving object takes is its

_____.

summer
June 21–Sept. 22

fall
Sept. 22–Dec. 21

winter
Dec. 21–March 20

spring
March 20–June 21

Read a Diagram

Which season is it in the Northern Hemisphere when the North Pole is tilted away from the Sun?

_____

LOG ON *Science in Motion* Watch how Earth revolves at www.macmillanmh.com

How does the Sun's apparent path change over the seasons?

The apparent path of the Sun is different in summer than in winter. In summer the Sun rises much higher in the sky. It also rises earlier and sets later. This is because Earth is tilted toward the Sun. In winter the Sun stays much lower in the sky.

At the Equator

Near the equator the Sun's apparent path changes very little during the year. Temperatures there change little from season to season. The way sunlight hits Earth's surface is almost the same all year long at the equator.

Apparent Path of the Sun

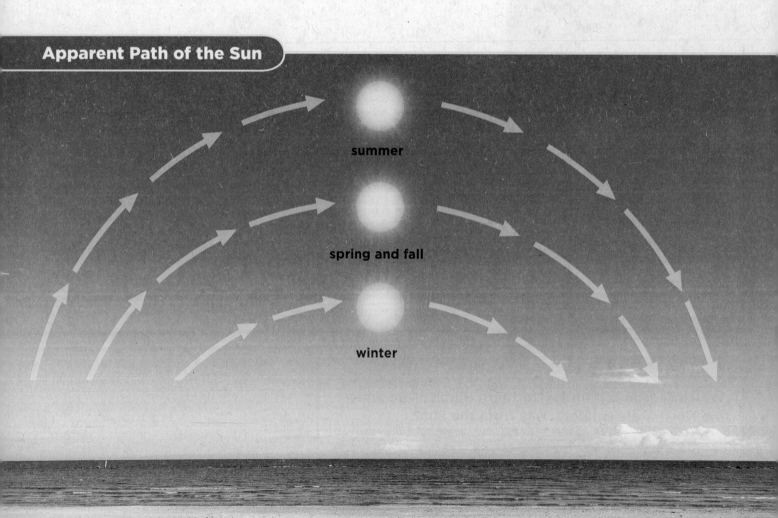

summer

spring and fall

winter

At the Poles

Near the poles the Sun's apparent path is very different between seasons. In northern Alaska, summer nights are very short. However, the Sun is not seen much during winter.

Making Predictions

The Sun's apparent path changes in the same way every year. Scientists use this pattern to predict when the Sun will rise and set.

▲ Antarctica is at the South Pole. Summer days are very long at the poles.

✔ Quick Check

Match the places on Earth to the Sun's apparent path.

6. ____ equator   **a.** Apparent path does not change very much all year.

7. ____ poles   **b.** Apparent path is very different between seasons.

What is the Moon like?

The Moon shines in the night sky. However, the Moon does not make its own light. Light from the Sun bounces off the Moon. Moonlight is reflected sunlight.

The Moon and Earth

The Moon is Earth's closest neighbor. Rocks on the Moon are like some rocks on Earth. However, the Moon and Earth are different in many ways.

The Moon is much smaller than Earth. It has no atmosphere. It has almost no water. The Moon is very hot during the day. The nights are colder than any place on Earth.

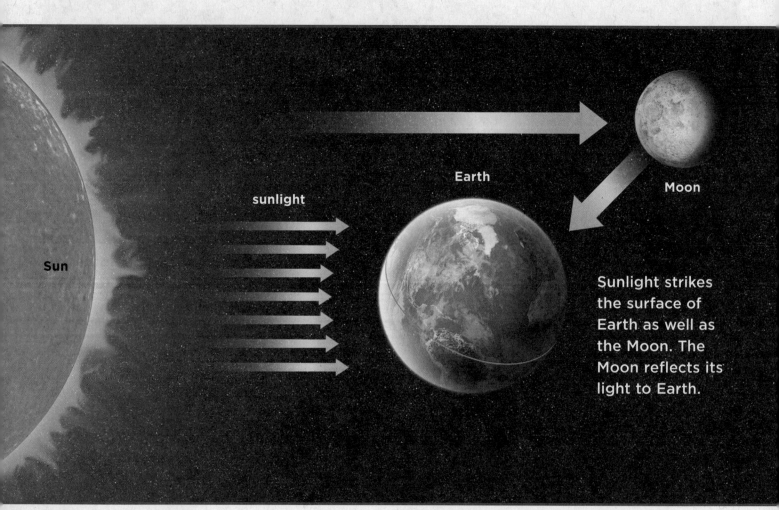

Sun

sunlight

Earth

Moon

Sunlight strikes the surface of Earth as well as the Moon. The Moon reflects its light to Earth.

crater

Surface Features

Most of the Moon's surface is covered with craters (KRAY•tuhrz). A **crater** is a hollow area or pit in the ground. Large rocks called meteoroids (MEE•tee•uh•roydz) crashed into the Moon. This made many of the Moon's craters. Meteroids often crash into other space objects.

Craters and Earth's Atmosphere

Earth's atmosphere keeps many meteoroids from crashing into Earth. When meteoroids enter Earth's atmosphere, they become very hot. Most burn up before they hit Earth's surface. This is why Earth's surface is not covered with craters.

 **Quick Check**

8. What is a crater?

_____

_____

What are the phases of the Moon?

The Moon revolves around Earth. It completes one orbit in just over 29 days.

The Moon's shape seems to change as it moves around Earth. The apparent shapes of the Moon are called **phases** (FAYZ•ez). During one complete orbit, the Moon goes through all of its phases.

Like Earth, half of the Moon is always lighted by the Sun. The other half is dark. We see different amounts of the lighted side during the Moon's orbit. Some moon phases include

• New Moon: none of the lighted side shows
• First Quarter Moon: half of the lighted side shows
• Full Moon: entire lighted side shows
• Third Quarter Moon: half of the lighted side shows

The Moon's Gravity

The Moon has gravity. It pulls a little on Earth. The Moon's gravity causes tides on Earth. Tides are the daily rise and fall of the ocean's surface.

✓ Quick Check

Fill in the blanks to complete the paragraph.

The **9.** _____ revolves around Earth. The

Moon completes one **10.** _____ around

Earth in just over 29 days. During one complete orbit,

the Moon goes through all of its **11.** _____.

During the **12.** _____ phase, you can see

the entire lighted side of the Moon.

Third Quarter Moon
The Moon is three quarters of the way around Earth.

Waning Crescent Moon
The left sliver of the Moon is the only part you can see.

Waning Gibbous Moon
Slightly less of the lighted side can be seen.

New Moon
The lighted side cannot be seen from Earth.

Full Moon
The entire lighted side can be seen.

Waxing Crescent Moon
Some of the lighted side can be seen.

Waxing Gibbous Moon
The Moon is almost full.

First Quarter Moon
The Moon is a quarter of the way around Earth.

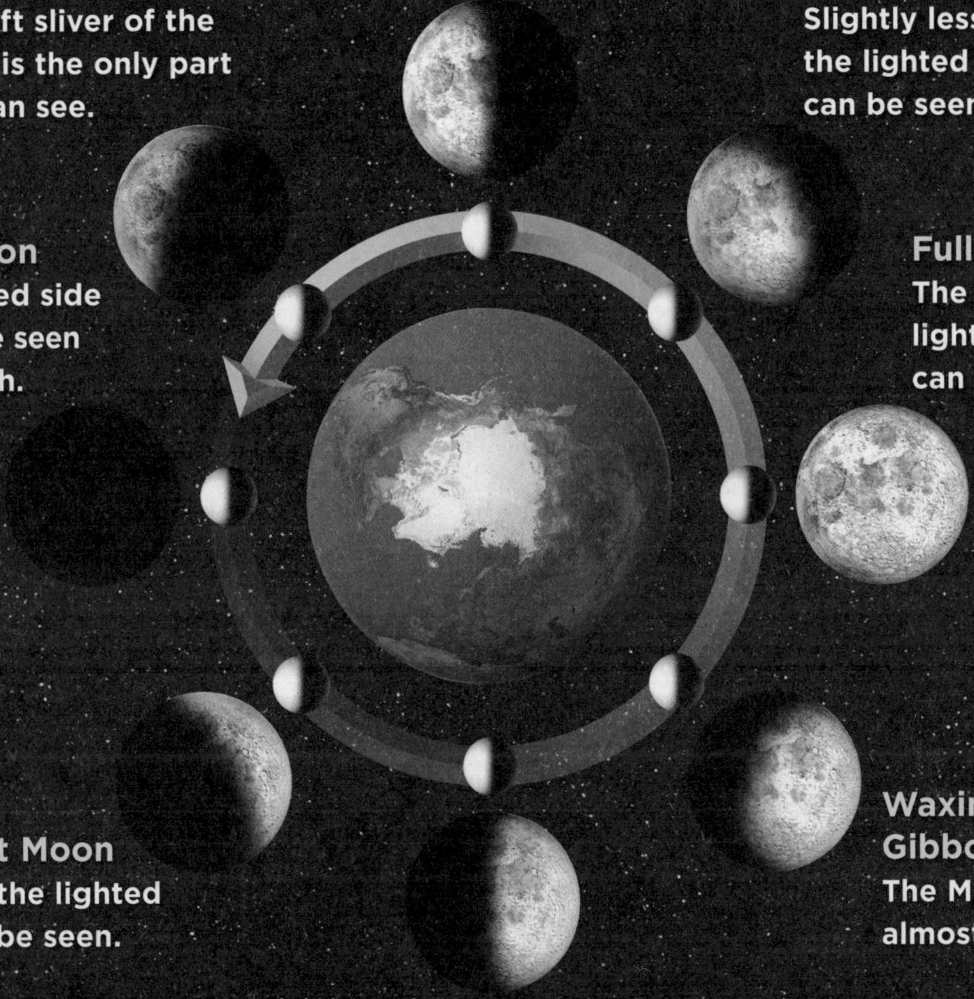

Read a Diagram

Which phase is before a full moon?

_____

What is an eclipse?

An eclipse (i•KLIPS) is a shadow cast by Earth or the Moon.

Solar Eclipses

In a **solar eclipse**, the Moon makes a shadow on Earth. This happens when the Moon is directly between the Sun and Earth. A part of Earth passes through the Moon's shadow. During a solar eclipse, the Moon appears to block out the Sun. Solar eclipses happen only during the new moon.

Two Kinds of Eclipses

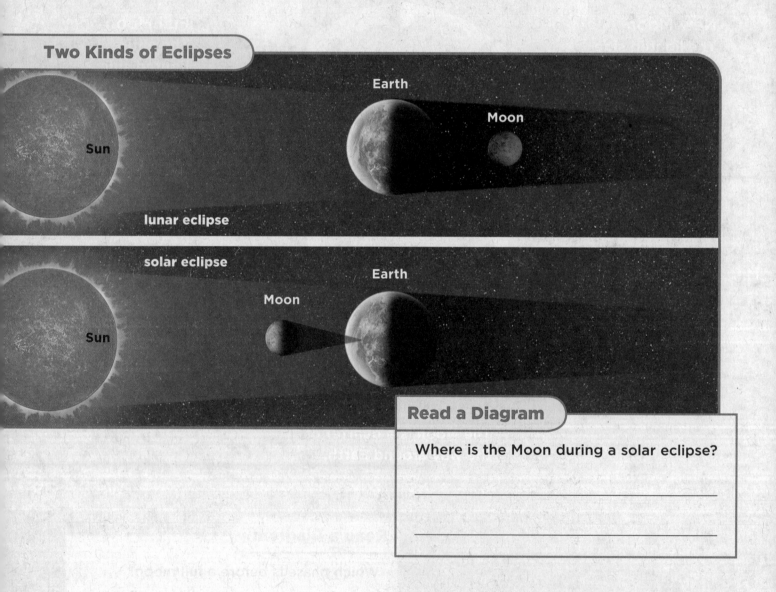

Sun

Earth

Moon

lunar eclipse

solar eclipse

Sun

Moon

Earth

Read a Diagram

Where is the Moon during a solar eclipse?

_____

_____

▲ Lunar eclipses can only happen at the full moon.

Lunar Eclipses

In a **lunar eclipse**, Earth makes a shadow on the Moon. This happens when Earth is directly between the Sun and the Moon. The Moon seems to disappear.

Eclipse Safety

You should look only at a lunar eclipse. Looking at a solar eclipse can hurt your eyes. Scientists use special tools to watch a solar eclipse safely.

✓ Quick Check

13. Where does the shadow fall during a lunar eclipse?

_____

14. Where does the shadow fall during a solar eclipse?

_____

What is the solar system?

The Moon is a satellite of Earth. A satellite is any object that moves in orbit around another, larger body. The Sun has many satellites. The Sun and its satellites make up our **solar system**. The Sun is the center of the solar system.

Planets

Planets are large, round objects that travel around the Sun. There are eight planets in our solar system. Planets are smaller and cooler than stars. Unlike a star, planets do not make light. However, planets reflect sunlight just like the Moon does. So you can see some planets in the night sky.

The Solar System

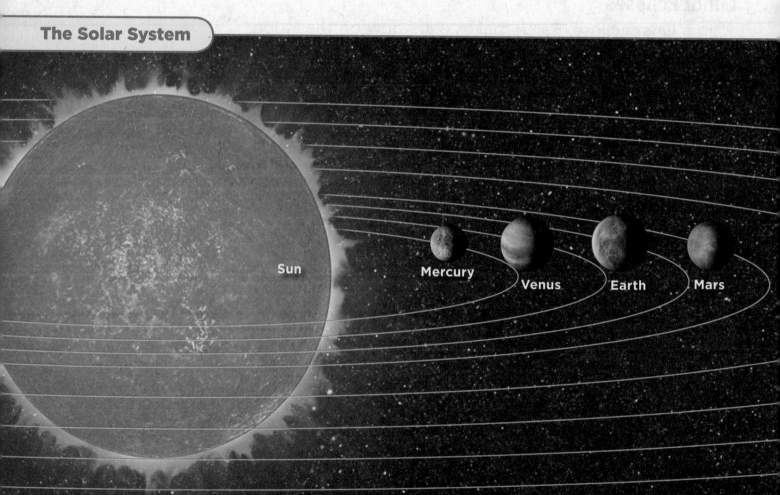

Sun

Mercury

Venus

Earth

Mars

Orbiting the Sun

The planets' orbits are ellipses (i•LIP•seez). An ellipse is a flattened circle, or oval.

Planets stay in orbit because of gravity and inertia (i•NER•shuh). **Gravity** is a force of attraction between all objects. It pulls planets to the Sun. Inertia is the tendency of a moving object to keep moving in a straight line. Together, gravity and inertia cause the planets to move around the Sun.

Quick Check

Write the definition of each of these words.

15. planet

_____

_____

16. gravity

_____

_____

Read a Diagram

Which planet has the longest journey around the Sun?

_____

Jupiter

Uranus

Neptune

Saturn

How do we learn about the solar system?

Galileo Galilei was an Italian scientist who studied the solar system. He put curved pieces of glass in a tube. This helped him look into space.

Telescopes

Galileo's tool was a telescope. **Telescopes** make objects that are far away seem closer. Galileo found objects in space that no one had seen before. Some of today's telescopes are like Galileo's, but larger. Other telescopes use curved mirrors.

Stars give off light. They also give off radio waves that we cannot see. Radio telescopes can read these waves.

Astronauts

In the 1960s, NASA—the National Aeronautic and Space Administration—used rockets to put people into space. Those people were the first astronauts (AS•truh•nawts).

New and Old Telescopes

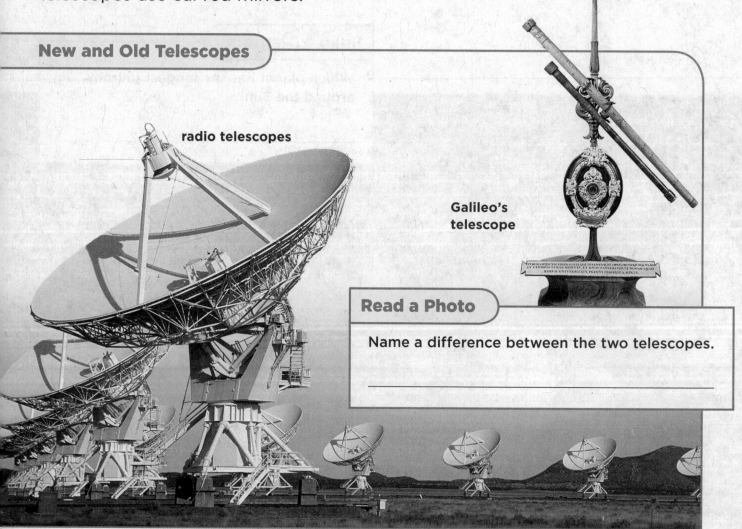

radio telescopes

Galileo's telescope

Read a Photo

Name a difference between the two telescopes.

_____

Shuttles and Space Stations

Space shuttles take astronauts into space to do experiments and put new satellites into space. Space shuttles can return to Earth. Space stations stay in space a long time. The International Space Station is used by many different countries.

Probes

A probe is a spacecraft that leaves Earth's orbit. It does not carry people. NASA has sent probes to planets, moons, and other objects. The probes are controlled by scientists on Earth. The probes send back pictures and data to Earth.

◀ In 2004 a space probe landed on Mars. Two robots, called Mars rovers, explored the surface of Mars.

✔ Quick Check

Name three tools scientists use to study space.

17. _____

18. _____

19. _____

What are the rocky planets?

The four planets closest to the Sun are called the rocky planets. They are made up of mostly rock. They also seem to have solid cores made of iron.

Mercury

Distance to the Sun: 58 million km
Diameter: 4,880 km
Rotation Time: 59 Earth days
Revolution Time: 88 Earth days
Fast Fact: Mercury's surface is covered with craters.

Mercury is the closest planet to the Sun. That makes it very hot. It is also the smallest planet. Mercury does not have a moon.

Venus

Distance to the Sun: 108 million km
Diameter: 12,100 km
Rotation Time: 243 Earth days
Revolution Time: 225 Earth days
Fast Fact: Temperatures on Venus can reach 500°C.

Venus is the second planet from the Sun. Venus has a very thick atmosphere. Its atmosphere holds in heat, making Venus the hottest planet. Venus also has many volcanoes.

Earth

Distance to the Sun: 150 million km
Diameter: 12,756 km
Rotation Time: 1 Earth day
Revolution Time: 365 Earth days
Fast Fact: Earth's atmosphere makes it suitable for life.

   Earth has an atmosphere and a surface of mostly liquid water. It is the only planet known to support life.

Mars

Distance to the Sun: 228 million km
Diameter: 6,794 km
Rotation Time: about 1 Earth day
Revolution Time: 687 Earth days
Fast Fact: Iron oxide, or rust, gives Mars its reddish color.

   The fourth planet from the Sun is Mars. Mars has two small moons and a thin atmosphere. Mars appears to have frozen water just below the surface.

✔ Quick Check

List the planets in order according to how close they are to the Sun.

Earth   Venus   Mars   Mercury

20. _____

21. _____

22. _____

23. _____

What are the other planets?

The four planets beyond Mars are called gas giants. They are large and made mostly of gases. They do not have solid surfaces.

Dwarf Planets

Scientists have found smaller planets in our solar system. They are called dwarf planets. Most are made of rock and ice. Pluto is a dwarf planet.

Jupiter

Distance to the Sun: 778 million km
Diameter: 143,000 km
Rotation Time: 10 Earth hours
Revolution Time: 4,333 Earth days
Fast Fact: Jupiter's four largest moons were first observed by Galileo in 1610.

Jupiter is the largest planet in the solar system. It has more than 60 moons. The surface is divided into bands. One band has a red spot the size of Earth.

Saturn

Distance to the Sun: 1 billion, 429 million km
Diameter: 120,536 km
Rotation Time: 10 Earth hours
Revolution Time: 10,759 Earth days
Fast Fact: Winds on Saturn can blow at 500 meters per second.

Saturn, the sixth planet from the Sun, is the second-largest planet. It has rings that circle the planet. Saturn has 34 known moons.

Uranus

Distance to the Sun: 2 billion, 871 million km

Diameter: 51,118 km

Rotation Time: 17 Earth hours

Revolution Time: 30,684 Earth days

Fast Fact: The axis of Uranus points toward the Sun.

Uranus is the only planet that rotates on its side. That means one pole always points toward the Sun. The other pole is always dark. Uranus has at least 27 moons.

Neptune

Distance to the Sun: 4 billion, 504 million km

Diameter: 49,528 km

Rotation Time: 16 Earth hours

Revolution Time: 60,190 Earth days

Fast Fact: Neptune takes 165 Earth years to orbit the Sun.

Neptune is the farthest planet from the Sun. It has 13 moons. Winds on the surface of Neptune blow up to 2,000 km (1,200 mi) per hour.

✓ Quick Check

Write the name of the planet next to the fact that describes it.

Jupiter   Saturn   Uranus   Neptune

24. _____   the farthest gas giant from the Sun

25. _____   has at least 34 moons

26. _____   rotates on its side

27. _____   largest gas giant

What else is in our solar system?

Planets are not the only objects in our solar system. Smaller objects also revolve around the Sun.

Comets
- A **comet** is a chunk of ice, rock, and dust that moves around the Sun.
- It moves in a long, narrow orbit.
- A comet heats up when it nears the Sun. This causes a tail to form.

Asteroids
- Asteroids (AS•tuh•roydz) are large pieces of rock or metal in space.
- Most asteroids are in a belt between Mars and Jupiter.

Comet Hale-Bopp last approached the Sun in the 1990s. ▼

▲ A meteoroid is called a meteor when it enters Earth's atmosphere.

Meteoroids
- smaller pieces of asteroids
- If a meteoroid enters Earth's atmosphere, it is called a meteor.
- Small meteors burn up in the atmosphere.
- Shooting stars are burning meteors.
- A meteor that reaches Earth is called a meteorite.

✓ Quick Check

Describe the three smaller objects found in our solar system.

28. _____

29. _____

30. _____

What are stars?

A **star** is a ball of hot gases that gives off light and heat. The only star you can see in the daytime is the Sun. The Sun is the closest star to Earth. This is why the Sun seems bigger and brighter than other stars. However, the Sun is an average size. Its temperature is average too.

Colors and Temperature

Stars are different colors. A star's temperature affects its color. The Sun's temperature makes it look yellow. Cooler stars are red or orange. Warmer stars are white or blue.

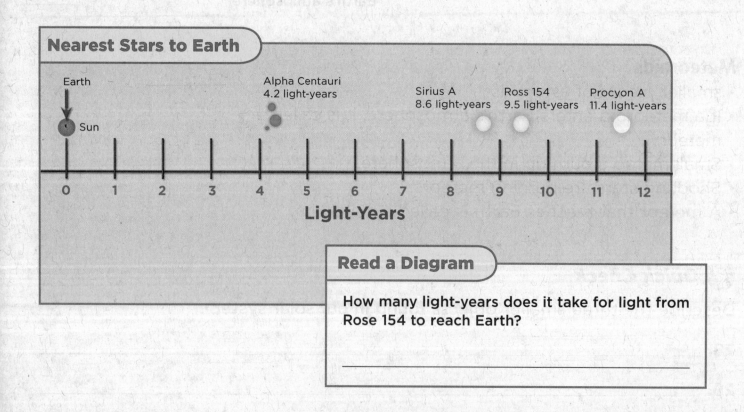

Nearest Stars to Earth

Earth

Sun

Alpha Centauri
4.2 light-years

Sirius A
8.6 light-years

Ross 154
9.5 light-years

Procyon A
11.4 light-years

0   1   2   3   4   5   6   7   8   9   10   11   12

Light-Years

Read a Diagram

How many light-years does it take for light from Rose 154 to reach Earth?

_____

▲ The Andromeda Galaxy is wider than our own Milky Way.

Light-Years

The Sun is 150 million kilometers from Earth. It takes about eight minutes for its light to reach Earth. Most stars are much farther away. They are so far away that scientists measure their distance in light-years. One light-year is the distance light travels in one year. This is nearly ten trillion kilometers.

Galaxies

All over the universe, stars are found in large groups called galaxies (GAL•uhk•seez). Our galaxy is the Milky Way. Our nearest neighbor is the Andromeda (an•DROM•i•duh) Galaxy. Both of these galaxies are shaped like spirals.

 ## Quick Check

Tell whether the sentences are true or false. If false, correct the sentence to make it true.

31. A star is a ball of hot gases that gives off light and heat.

_____

32. Solar systems are large groups of stars.

_____

What are constellations?

There are billions of stars. It is hard to make sense of all of them. One way is to group them into constellations (kon•stuh•LAY•shuhnz). A **constellation** is a group of stars that make a pattern or picture in the sky. People named constellations after pictures they saw in the sky.

The constellations you see depend on where you are on Earth. The night sky is different in different places.

The night sky also looks different in winter than it does in summer. As Earth travels around the Sun, we see different constellations. Because of this, you can use constellations to tell what season it is.

Constellations

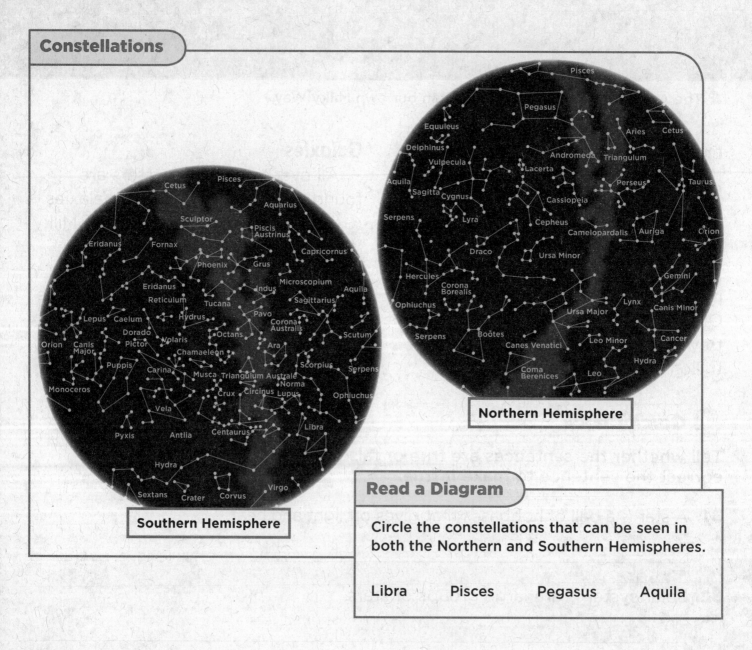

Southern Hemisphere

Northern Hemisphere

Read a Diagram

Circle the constellations that can be seen in both the Northern and Southern Hemispheres.

Libra   Pisces   Pegasus   Aquila

Marking Time and Seasons

Once there were no clocks to tell time. Instead, people used constellations.

Farmers used constellations to mark the seasons. The stars' positions helped them know when to plant crops. Sailors used constellations to steer their ships. They knew which stars were in the northern sky.

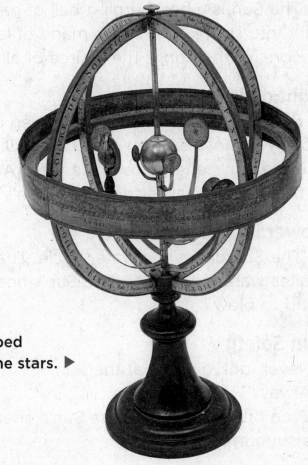

This ancient tool helped people tell time by the stars. ▶

 **Quick Check**

33. What is a constellation?

_____

_____

34. How did people in the past use constellations?

_____

_____

What is the Sun like?

The Sun is a hot, burning ball of gas. It gives off light into space. The Sun is made of layers. The center, or core, of the Sun is the source of all its energy.

Light and Heat Energy

Some of the Sun's energy is given off as light. However, most is given off as heat. On Earth, plants use the Sun's energy to make food. Animals eat plants. In this way animals also use the Sun's energy.

Power for the Water Cycle

The Sun drives the water cycle. The Sun's heat makes water evaporate. The Sun's heat also causes wind to blow.

Sun Safety

- Never look directly at the Sun.
- Always wear sunscreen.
- Even on a cloudy day, the Sun's energy can give you a sunburn.

▼ Here you see parts of the Sun that you cannot see from Earth.

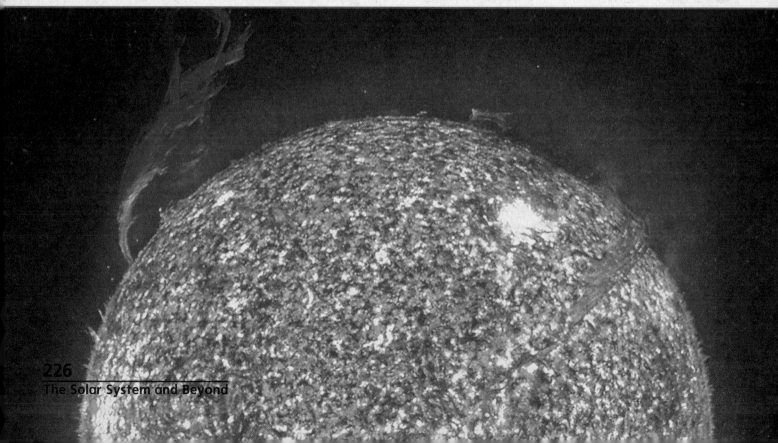

▲ The Sun heats Earth's water and air. This causes water to evaporate and wind to blow.

✓ Quick Check

Tell whether the sentences are true or false. If false, correct the sentence to make it true.

35. The Sun gives off light into space.

_____

36. You do not need to wear sunscreen on a cloudy day.

_____

37. The Sun drives the water cycle.

_____

The Solar System and Beyond

Use the clues below to fill in the crossword puzzle.

Across

1. a hollow area or pit in the ground

3. one complete trip around an object in a circular or nearly circular path

4. the spinning of an object around its axis

5. Earth casting a shadow on the Moon

8. a round object in space that is a satellite of the Sun

Down

1. a group of stars that form a pattern in the night sky

2. the Moon casting a shadow on Earth

6. a real or imaginary line that an object spins around

7. a sphere of hot gases that gives off light and heat

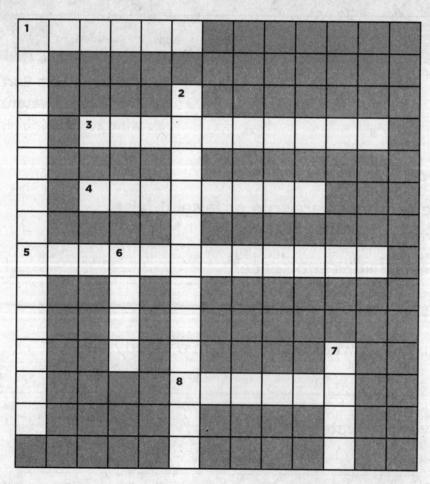

| comet | gravity | planet | solar system |
| constellation | phase | rotation | telescope |

Use each word just once to fill in the blanks.

1. A tool that makes objects look closer and larger is

    called a(n) _____.

2. A force of attraction between all objects is called

    _____.

3. A group of stars that appear to form a pattern in the

    night sky is called a(n) _____.

4. A chunk of ice, rock, and dust that moves around the

    Sun is called a(n) _____.

5. An apparent change in the Moon's shape is called

    a(n) _____.

6. A round object in space that is a satellite of the Sun is

    called a(n) _____.

7. The complete spin of an object around its axis is

    called a(n) _____.

8. The Sun and all the objects that travel around it is

    the _____.

Summarize

_____

_____

_____

CHAPTER 9

Properties of Matter

Vocabulary

 **matter** anything that has mass and takes up space

 **property** a characteristic of matter you can observe

 **mass** the amount of matter in an object

 **volume** the amount of space an object takes up

 **solid** matter with a definite shape that takes up a definite volume

 **liquid** matter that has a definite volume but no definite shape

 **gas** matter that has no definite shape or volume

What is matter and how is it classified?

length the number of units along the edge of an object

density the amount of matter in a certain amount of space

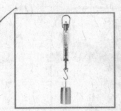
weight the measure of the pull of gravity

element a substance that is made up of only one type of matter

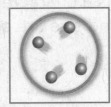
atom the smallest part of an element

metal an element that is shiny and conducts heat and electricity

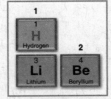
periodic table a chart that shows the elements classified by their properties

What is matter?

Almost everything around you is matter. **Matter** is anything that has mass and takes up space. This book is matter. Light and heat are not matter. They do not take up space.

You can describe matter by its properties (PROP•uhr•teez). A **property** is a characteristic that you can observe. Color, shape, and size are some properties of matter.

Matter Has Mass

One property of matter is mass. **Mass** is the amount of matter in an object. A balance (BAL•uhns) is a tool used to measure mass. Mass is often measured in grams (g) or kilograms (kg).

Comparing Masses

Read a Photo

Look at the balance. How can you tell that the rock has more mass?

_____

_____

Matter Has Volume

Volume (VOL•yewm) is also a property of matter. **Volume** is how much space an object takes up. We measure volume by counting the number of cubes that fit in an object. We measure volume by using a measuring cup.

Useful Properties

Properties help us choose the right kind of matter for different jobs. We use iron when we need strength. We use wood when we need matter that can be cut and shaped. Some matter can float in water, such as cork and wood.

Some properties of matter cannot be seen. However, these properties are still useful. For example, some metals are attracted to a magnet. Some matter burns easily and gives off heat. Some matter does not burn.

Magnetism is a property of matter. ▶

▲ Salt dissolves in water.

Sand does not dissolve in water. ▼

◀ Some objects can float in water. Other objects sink.

✔ Quick Check

Fill in the blanks with the correct word.

1. The amount of matter making up an object is its _____.

2. Anything that has mass and takes up space is _____.

3. Color is a _____ of matter.

What are the states of matter?

Matter comes in many forms. We call these forms states. Solid, liquid, and gas are the three common states of matter.

Solid

A **solid** has a definite shape and takes up a definite volume. The particles of a solid are close together. They do not move around easily.

Liquid

A **liquid** has a definite volume but no definite shape. A liquid takes the shape of its container. The particles of a liquid are not as close together as a solid's particles are. The particles can move past each other.

Gas

A **gas** has no definite shape or volume. A gas fills the shape and the volume of its container. The particles of a gas move around easily. If a balloon breaks, the gas inside it will spread out into the air.

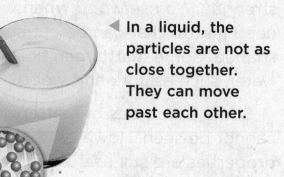

In a solid, the particles are packed closely together.

In a liquid, the particles are not as close together. They can move past each other.

▲ In a gas, the particles are far apart. Gas fills the shape of the balloon.

✅ Quick Check

In each row, cross out the word that does not belong.

4. solid   log   book   juice

5. liquid   milk   juice   air

6. gas   log   air   fills a space

What happens to the matter we use?

You use matter all the time. Some matter, such as air, can be used again and again. Other forms of matter get thrown away. Matter can become trash. Trash goes into landfills or oceans.

You can reuse matter. An egg carton can be reused. You can plant seeds in the cups.

Matter can be recycled, or made into something else. Cans, paper, plastic, and glass can be recycled.

✓ Quick Check

Write true or false in the blank.

7. Matter can be reused.

_____

8. Matter can be recycled.

_____

9. Matter can become trash.

_____

10. Matter cannot be used again and again. _____

Uses of Matter

Objects Made by People

Objects in Nature

Read a Photo

Matter can be sorted into groups. What are the two groups of matter in these photos?

_____

_____

How do we measure matter?

In the United States, we measure with standard units. Inches (in.), pounds (lbs), and ounces (oz) are all standard units.

Scientists use the metric system. The metric system is based on units of 10. It uses prefixes, such as centi– and milli–. One meter has 100 centimeters. One kilometer has 1,000 meters.

Metric Units	Amount	Estimated Length
1 centimeter (cm)	$\frac{1}{100}$ of a meter	the width of your thumbnail
1 meter (m)	100 cm	the length of a baseball bat
1 kilometer (km)	1,000 m 100,000 cm	the distance you walk in 10 to 15 minutes

Read a Table

Which is longer—a kilometer or a meter?

_____

Length and Width

An object's **length** is the number of units along the edge of an object. Width is the number of units across an object. You can measure length and width with a ruler.

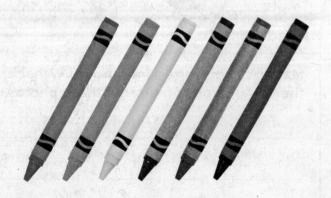

◀ You can measure length in centimeters or inches.

Area

Area (AYR•ee•uh) tells the number of squares that cover a surface. To find area, multiply length times width. The length of a rug might be 3 meters. The width of the rug might be 2 meters. The area of this rug is 3 meters × 2 meters, or 6 square meters.

Volume

Volume is the amount of space an object takes up. You can use math to find the volume of a box. Multiply the box's length by the width by the height. You can also measure volume with a measuring cup, a spoon, a graduated cylinder, or a beaker.

A baker may measure volume in cups or pints. ▼

A measuring spoon can measure volume. ▶

Quick Check

Draw a line from the type of measurement to the best way to find that measurement.

11. area   **a.** multiply length times width

12. volume   **b.** use a ruler

13. length   **c.** use a measuring cup

What is density?

A plastic ball floats on water. If you fill the ball with sand, it will sink. The volume of the ball is the same. However, the ball filled with sand has more mass.

You can also say that the ball filled with sand has greater density. **Density** is the amount of matter in a given space. The ball with sand has more matter than the ball without sand.

Density describes the relationship between mass and volume. To find density, we divide mass by volume. Remember that mass is the amount of matter in an object. Volume is how much space an object takes up.

The particles of cork are loosely packed together. Cork is not very dense. ▶

The particles that make up marble are packed closer together. Marble is denser than cork. ▶

A brass key is very dense. The particles are tightly packed together. ▶

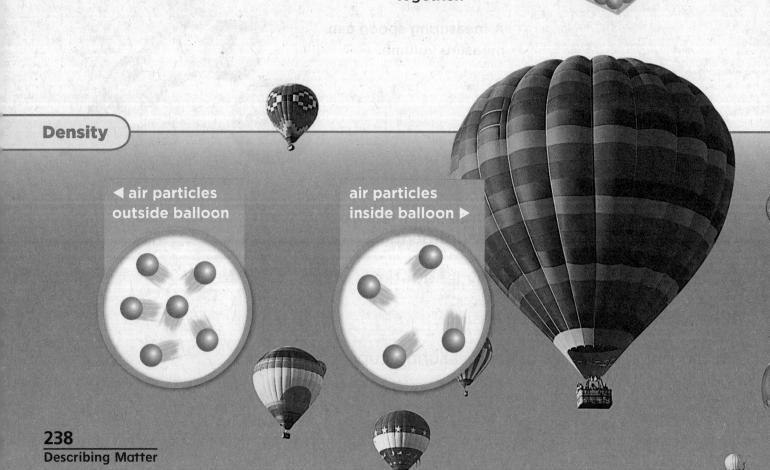

◀ air particles outside balloon

air particles inside balloon ▶

Float or Sink?

The density of cork is less than the density of water. Cork will float on water. The density of a brass key is greater than the density of water. A brass key will sink.

You can change the density of matter. If you add heat to air, the air particles move around more. They spread out. Heated air is less dense than cooler air. Heated air floats over denser, cooler air. This is how a hot air balloon rises into the air.

 Quick Check

Circle the correct answer.

14. The density of cork is less than the density of water.

   true   false

15. Density tells how tightly matter is packed together.

   true   false

16. Cooler air rises above heated air.

   true   false

Read a Diagram

How do you know that the air inside the balloon is less dense than the air outside the balloon?

_____

_____

_____

What is weight?

Weight (WAYT) is another way to measure matter. **Weight** measures the pull of gravity on an object. Gravity is a force, or pull, between all objects. The stronger the pull of gravity, the more an object weighs.

Gravity depends on mass. An object with more mass has a stronger pull of gravity. Earth has more mass than the Moon. Earth has more gravity. The same object will weigh more on Earth than it will on the Moon. However, an object will have the same mass on Earth and the Moon.

Earth

▲ On both Earth and the Moon, the object has a mass of 1 kg. On Earth, the object's weight is 9.8 newtons.

Mass and weight are not the same. Remember that mass is the measure of how much matter is in an object. Mass is measured with a balance. Weight is measured with a scale.

The English units for weight are pounds and ounces. The metric unit for weight is the newton (N).

Moon

▲ An object that weighs 9.8 newtons on Earth will weigh only 1.6 newtons on the Moon.

✓ Quick Check

17. Will a rock weigh more or less on the Moon? Explain your answer.

_____

_____

_____

_____

_____

What are elements?

All matter is made of elements (EL•uh•muhnts). An **element** is a substance that is made up of only one type of matter. An element cannot be broken down into a simpler form.

Hydrogen is an element. Oxygen, gold, and silver are elements, too. Elements are like building blocks. They can be used to make different substances. Hydrogen and oxygen are the building blocks of water.

Atoms

Elements are made up of atoms (AT•uhmz). An **atom** is the smallest part of an element. Atoms are tiny. You cannot see them.

All the atoms in an element are alike. Copper is an element. Copper atoms are the same. Copper atoms are different from gold atoms.

Neon is an element. It is a gas. In a tube, neon can glow.

Metals and Nonmetals

Elements are organized into three groups. A **metal** (MET•uhl) is shiny. It can be bent into a shape. Most metals let heat and electricity pass through them easily. Two metal elements are iron and copper.

Most nonmetals have different properties from metals. Most nonmetals do not let heat and electricity pass through them easily. Many are gases. Oxygen and carbon are nonmetals.

Metalloids (MET•uh•loydz) have some of the properties of both metals and nonmetals. Silicon is a metalloid.

Symbols for Elements

Each element has a symbol. Carbon is an element. The symbol for carbon is the letter C. Symbols make it easier for scientists to write chemical formulas.

 Quick Check

List three details that support the main idea.

Main Idea	Details
There are three groups of elements.	18. _____
	19. _____
	20. _____

▲ Aluminum is a strong metal. It is light in weight.

pure copper

▲ Copper is used to make jewelry.

Periodic Table of the Elements

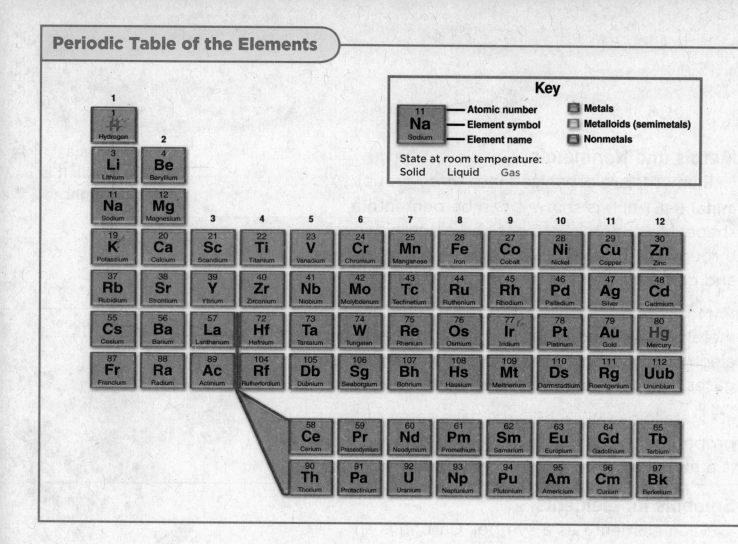

Key

11 | Na | Sodium — Atomic number / Element symbol / Element name

Metals
Metalloids (semimetals)
Nonmetals

State at room temperature:
Solid   Liquid   Gas

How are the elements organized?

Elements are organized on the **periodic**
(peer•ee•OD•ik) **table**. This table organizes the elements
by their properties. Elements are listed in order by their
atomic number. The atomic number is related to the
mass of each element.

Elements are placed in rows and columns. Elements in
the same column have properties in common. For
example, all the elements in column 17 combine easily
with other elements.

Elements in the same row can also have properties in
common. Platinum (Pt) and gold (Au) are next to each
other. They are both shiny metals that are used in jewelry.

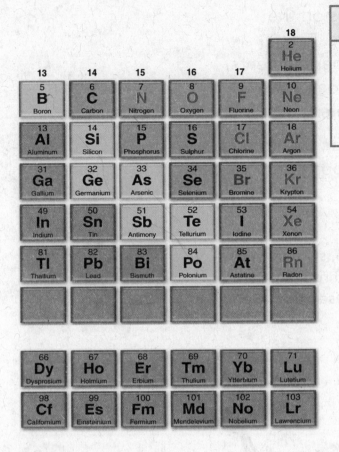

Read a Table

Are there more metals or nonmetals on the periodic table?

_____

Diamonds, coal, and the graphite in pencils are all made of carbon. ▶

✓ **Quick Check**

Classify these elements. Put a check in the correct column.

	Metal	Nonmetal	Metalloid
21. hydrogen (H)			
22. boron (B)			
23. carbon (C)			
24. helium (He)			
25. silicon (Si)			
26. potassium (K)			

How do scientists use the periodic table?

Scientists can look at where an element is on the periodic table and tell you how it will behave. For example, hydrogen combines easily with other elements. The elements in the same column with hydrogen will also likely combine easily with other elements.

Elements in the same row are often similar, too. Iron is magnetic. The two elements next to iron (cobalt and nickel) are also magnetic.

Chromium is used to make steel goods, such as bicycle frames. Manganese is next to chromium on the periodic table. Manganese is also used to make steel goods. ▶

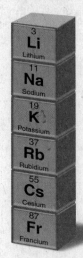

A banana has potassium (K). Elements in the potassium column react the same way with nonmetals.

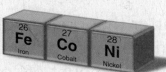

Nails can be made with iron (Fe), cobalt (Co), or nickel (Ni). These elements are all magnetic.

This mineral contains fluorine (F). Elements in the fluorine column make salts when combined with elements in column 1.

Look back at the periodic table. You can see that some parts of the table are not filled in all the way. Why not? Scientists might not have found all the elements. New elements could possibly exist.

 Quick Check

27. List three magnetic elements.

_____

_____

_____

28. What is similar about their position on the periodic table?

_____

_____

_____

Read a Diagram

Do you think sodium (Na) will react with nonmetals? Explain your answer.

_____

_____

Properties of Matter

Choose the letter of the best answer.

1. Matter that takes the shape of its container is a
- **a.** solid.
- **b.** liquid.
- **c.** gas.
- **d.** mass.

2. A characteristic of matter you can observe, such as color, shape, and size, is called a(n)
- **a.** property.
- **b.** element.
- **c.** periodic table.
- **d.** metric system.

3. The smallest part of an element is a(n)
- **a.** matter.
- **b.** metal.
- **c.** atom.
- **d.** gas.

4. The measure of the pull of gravity between an object and a planet, such as Earth, is called
- **a.** length.
- **b.** weight.
- **c.** volume.
- **d.** periodic table.

5. Matter with a definite shape that takes up a definite amount of space is
- **a.** solid.
- **b.** liquid.
- **c.** gas.
- **d.** length.

6. Matter that does not have a definite shape or take up a definite amount of space is a
- **a.** solid.
- **b.** liquid.
- **c.** gas.
- **d.** metal.

7. The amount of space an object takes up is its
- **a.** weight.
- **b.** mass.
- **c.** density.
- **d.** volume.

8. The amount of matter in a given space is
- **a.** volume.
- **b.** mass.
- **c.** length.
- **d.** density.

P	T	M	E	T	A	L	I	P
M	E	L	E	M	E	N	T	S
A	T	R	C	Y	O	K	A	M
T	V	A	I	G	T	I	D	A
T	A	I	B	O	Y	S	I	S
E	M	H	W	L	D	T	M	S
R	V	O	L	U	E	I	S	N
L	E	N	G	T	H	R	C	E

element ✓   length ✓   mass ✓   matter ✓   metal ✓   periodic table ✓

Fill in the blanks with words from the box. Then find each word in the puzzle.

1. Anything that has mass and takes up space is  matter .

2. A substance that is made up of only one type of matter is a(n)

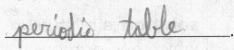 element .

3. A chart that shows the elements classified by properties is the

 periodic table .

4. The amount of matter making up an object is its

 mass .

5. The number of units that fit along one edge of an object is the object's

 length .

6. An element that is shiny and conducts heat and electricity is a(n)

metal .

Summarize

_____

_____

_____

CHAPTER 10

Matter and Its Changes

Vocabulary

physical change a change that begins and ends with the same type of matter

change of state a physical change of matter from one state to another

chemical change a change that results in matter that has different properties than the original matter

mixture two or more substances that are combined but that keep their original properties

solution a mixture in which one or more substances are blended evenly in another substance

alloy a solution in which at least one kind of matter is a metal

How can matter change?

 **filter** a tool that separates things by size

acid a substance that turns blue litmus paper red

 **compound** a substance made when two or more elements are joined together and lose their own properties

 **base** a substance that turns red litmus paper blue

 **distillation** a way to separate the parts of a solution

What are physical changes?

When you push down on a piece of clay, you cause a physical (FIZ•i•kuhl) change. The clay has a different shape, but it is still clay. A **physical change** begins and ends with the same type of matter. When you fold, cut, break, tear, bend, or stretch matter, you cause physical changes.

Heating and cooling can cause a special kind of physical change. Ice is frozen water. It is in the solid state. When you add heat, the ice melts into liquid water. The water is now in the liquid state. When liquid water boils, it turns into steam. Steam is a sign that the liquid water turned into water vapor. Water vapor is in the form of a gas. Freezing water is also a physical change.

▲ When you knit, you cause a physical change. The yarn changes shape, but the yarn is still yarn.

When you fold paper, you cause a physical change. ▼

Liquid water turns into a gas when it is heated. ▼

Real-World Changes

Physical changes happen all around you. Sidewalks are made of concrete. When the concrete is new, it is one big piece. In time, the concrete will chip and crack. Small bits will break off. The cracking and breaking are physical changes.

You may also see physical changes in winter. When it is cold, surfaces of lakes and ponds can freeze. The surfaces become ice. Under the ice, the water is still liquid. Fish can still swim in the lake.

Signs of a Physical Change

Some physical changes are hard to see. Look for a change in size, shape, or state. Even a change in position can be a sign of a physical change.

 **Quick Check**

Fill in the effect for each cause.

1. Cause: Pressing down on a piece of clay

Effect: _____

_____

2. Cause: Adding heat to ice

Effect: _____

_____

Water can wear away rock. This is a physical change.

How does matter change state?

Matter can exist in three states: solid, liquid, and gas. A **change of state** is a physical change when one state of matter changes into another. Matter changes state when energy is added or taken away.

Heating

When you heat a solid, you add energy to the solid. The solid's particles move faster. The particles begin to spread out. The solid melts. Melting is a change from a solid state to a liquid state.

If you add enough energy to a liquid, it can boil. Boiling is a change from a liquid to a gas. Boiling is not the only way a liquid can become a gas.

How Water Changes State

solid

Ice melts when energy is added. The particles move faster.

liquid

As energy is added to liquid water, the particles move even faster.

Evaporation

The Sun's energy evaporates water in ponds, lakes, rivers, and oceans. Evaporation (i•vap•uh•RAY•shuhn) is the slow change of a liquid to a gas without boiling.

Cooling

When matter loses energy, the particles move more slowly. In most kinds of matter, the particles also move closer together. This is called cooling. Gas condenses to a liquid when energy is lost. A liquid can freeze into a solid if it continues to cool.

When liquid water cools, its particles move closer together. When the temperature reaches 4°C, the particles move apart.

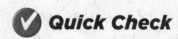

Water vapor is a gas. Its particles move very fast.

✔ **Quick Check**

3. When you add energy to a solid, it will change into a(n)

_____.

4. When you add energy to a liquid, it will change into a(n)

_____.

5. When you take away energy, gas will condense into a(n)

_____.

Read a Diagram

Why do the particles move faster in each picture?

_____

_____

What are chemical changes?

If you leave your bicycle out in the rain, it can rust. Iron in your bicycle is a solid metal. Iron can react with water and oxygen in the air to make rust. Rust is brown in color.

The change from iron to rust is a chemical change. A **chemical change** begins with one kind of matter and ends with another kind. The new matter has different properties from the original matter. Rust is very different from iron and oxygen.

Examples of Chemical Changes

Cooking and baking cause many chemical changes in food. A cooked egg has different properties from a raw egg. Bread has different properties from the yeast and flour that you start with.

Adding baking soda to vinegar causes another chemical change. A gas is let off. This bubbly gas is called carbon dioxide.

Gas bubbles and a fizzing sound
▼ are signs of a chemical change.

An explosion and glowing colors
▼ are signs of a chemical change.

Signs of a Chemical Change

You can see, smell, feel, and hear chemical changes.

One sign of a chemical change is a change in color. You can see a change in color when iron rusts.

You can smell some chemical changes. When you toast a marshmallow over a fire, it smells good!

A chemical change, such as wood burning, can give off heat and light. You can feel the fire's heat.

Some tablets fizz in water. A chemical change can make bubbles and a fizzing sound.

✓ Quick Check

Classify the following actions. Tell whether they are chemical changes or physical changes.

- cutting paper
- toasting a marshmallow
- melting ice
- rusting iron
- bending clay
- fizzing bubbles

Reaction of Iron and Sulfur

① Iron and sulfur are mixed together. Iron is a gray metal. It is also magnetic. Sulfur is a yellow powder.

② A metal rod is heated to a high temperature.

③ The heated rod causes a chemical change. Light and heat are released.

④ The result is iron sulfide—a black, nonmagnetic material.

Read a Diagram

What causes the chemical change?

_____

Chemical Change	Physical Change
6. _____	9. _____
7. _____	10. _____
8. _____	11. _____

What are other real-world changes?

Physical changes and chemical changes happen all the time. They happen all around you. The chart shows some examples.

Physical Changes

Outside Water vapor cools. It condenses into clouds. When the condensed drops get big enough, the drops fall as rain.

In Your Body Our bodies make sweat. The sweat can evaporate off our skin. This makes us feel cooler.

In the Kitchen Dough is easy to pull and push. Dough can be shaped into rolls.

Chemical Changes

Outside Some rain falls as acid rain. Acid rain reacts with limestone. Acid rain can eat away the stone.

In Your Body Blood carries oxygen to your cells. In the cell, the oxygen reacts with sugars. This releases energy for your body.

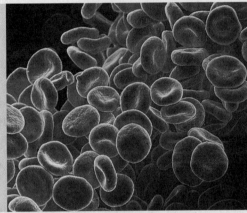

In the Kitchen Heat from an oven can bake the dough into bread. The dough was soft and moist. The bread crust is firm and dry.

✔ Quick Check

12. Describe a chemical change and a physical change that happens to bread dough.

Chemical change: _____

Physical change: _____

What is a mixture?

A salad is a mixture. So is a bowl of cereal. A **mixture** is a combination of two or more kinds of matter. In a mixture, each kind of matter stays the same.

Everyday Mixtures

A salad is a mixture of foods, such as lettuce and carrots. In the salad, a carrot still tastes like a carrot. It is still orange. It keeps its carrot properties.

Some cereals are a mixture of solids, such as grains and nuts. You can add milk to cereal. Then you have a mixture of solids and a liquid.

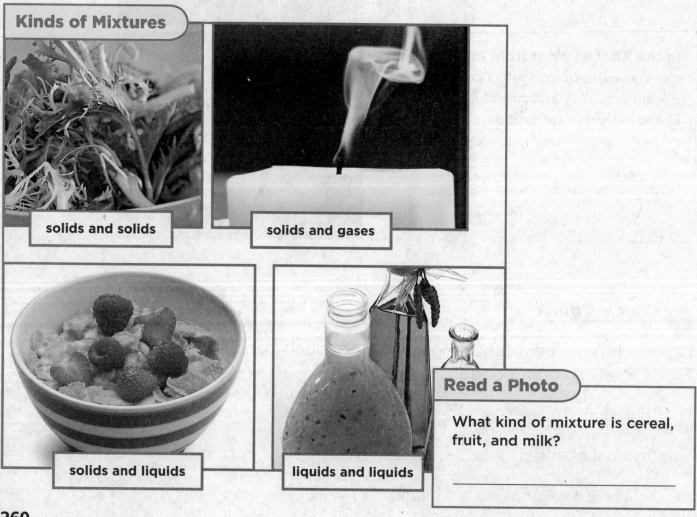

Kinds of Mixtures

solids and solids

solids and gases

solids and liquids

liquids and liquids

Read a Photo

What kind of mixture is cereal, fruit, and milk?

_____

Solutions Are Mixtures

Some solids, such as salt, mix easily with liquids. If you mix salt in water, you cannot see the salt. The salt and the water have become a solution. A **solution** is a mixture of two or more substances that blend together evenly.

Alloys Are Solutions

A special kind of solution is an alloy. An **alloy** is a mixture of two or more elements. At least one element in an alloy is a metal. Bronze is an alloy. It is a solution of copper and tin. Steel is an alloy of iron and carbon.

Lemonade is a solution of water, lemon juice, and sugar. ▶

▼ This brooch is made of bronze— an alloy of copper and tin.

✓ *Quick Check*

Write true or false for each statement.

13. A salt-and-water solution is a mixture. _____

14. An alloy is a solution made from metals. _____

15. A salad is a solution made from vegetables. _____

How can you separate the parts of a mixture?

The different parts of a mixture can be separated. For example, you can separate a mixture of beads and coins by their shapes and colors.

Settling

Another way to separate a mixture is by settling. Parts of a mixture settle when they have different densities. Mud is a mixture of soil and water. Over time, the soil in mud sinks to the bottom. Soil settles because it is denser than water.

Filtration

A **filter** separates things by size. A filter has holes. It may be a screen, a net, or a sieve (SIV). Matter that is smaller than the holes can pass through. Larger matter will stay behind. A colander is a kitchen tool. You can use a colander to separate noodles from water.

Magnets

You can use a magnet to separate parts of some mixtures. Magnets pull, or attract, iron, nickel, and cobalt. This property is called magnetic attraction.

Natural Settling

Read a Photo

Did soil get on this car through filtration, settling, or magnetic attraction?

_____

✓ Quick Check

16. Iron can be separated from a mixture with a(n)

_____ .

17. A colander is a(n) _____ because it can separate noodles from water.

How can you separate the parts of a solution?

Filters, sieves, or settling might not separate the parts of a solution. To separate them, you can use distillation and evaporation.

Distillation

In **distillation**, a solution is heated until the liquid turns into a gas. The solid part of the solution is left behind. The gas then passes through a tool that cools it. The gas condenses into a liquid.

Distillation is used to make fuel for our cars. The process separates gasoline from crude oil.

Evaporation

Evaporation is the change of a liquid into a gas without boiling. When salt water evaporates, the water becomes a gas. The solid salt is left behind. Evaporation collects only the solids in a solution. The liquids are lost to the air.

✓ Quick Check

18. What are two ways to separate the parts of a solution?

_____

_____

Distillation towers ▶ at this refinery separate gasoline from crude oil.

What are compounds?

Table salt is made of two elements: sodium and chlorine. Salt is a compound. A **compound** is made up of two or more elements that are joined together chemically. Unlike the parts of a mixture, the parts of a compound cannot be separated physically. A compound can be separated only by chemical means.

Chemical Properties

The properties of a compound are different from the properties of its original elements. When two or more elements join chemically, they lose their physical properties.

Rust is a compound made of two elements: iron and oxygen. Iron is a strong metal. Iron and oxygen combine to make rust. Rust is weaker than iron.

Comparing Compounds and Mixtures

	Compound	Mixture
How are the parts combined?	two or more elements are combined chemically	two or more types of matter are mixed together
Do the parts keep their own properties?	no	yes
How can it be separated?	by chemical means	by physical means

 Quick Check

19. How are the elements in a compound combined?

_____

20. Do the different parts of a compound keep their own properties? _____

Combining Elements

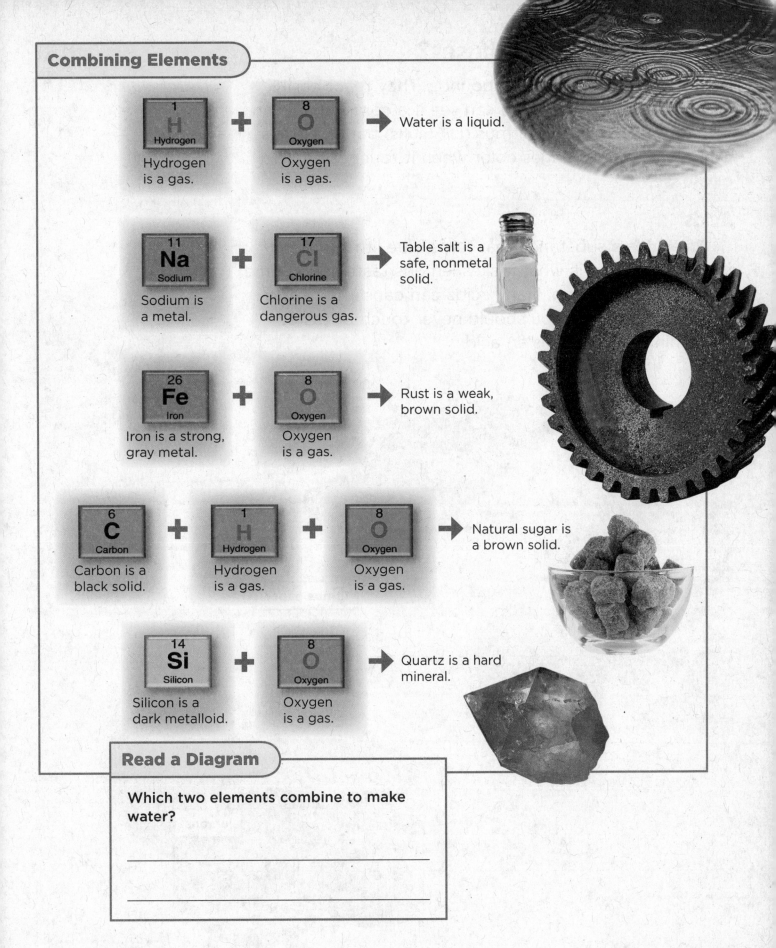

1 H Hydrogen	**+**	**8** O Oxygen
Hydrogen is a gas.		Oxygen is a gas.

→ Water is a liquid.

11 Na Sodium	**+**	**17** Cl Chlorine
Sodium is a metal.		Chlorine is a dangerous gas.

→ Table salt is a safe, nonmetal solid.

26 Fe Iron	**+**	**8** O Oxygen
Iron is a strong, gray metal.		Oxygen is a gas.

→ Rust is a weak, brown solid.

6 C Carbon	**+**	**1** H Hydrogen	**+**	**8** O Oxygen
Carbon is a black solid.		Hydrogen is a gas.		Oxygen is a gas.

→ Natural sugar is a brown solid.

14 Si Silicon	**+**	**8** O Oxygen
Silicon is a dark metalloid.		Oxygen is a gas.

→ Quartz is a hard mineral.

Read a Diagram

Which two elements combine to make water?

_____

_____

What are acids and bases?

Acids and bases are compounds. They react easily with other substances. You can see if a compound is an acid or a base by using litmus (LIT•muhs) paper. Litmus is a dye. Litmus changes color when it touches an acid or a base.

Acids

An **acid** is a substance that turns blue litmus paper red. A weak acid is what makes lemons taste sour. Some acids are very strong. Many acids can cause harm and can burn your skin. You should never touch or taste something to see if it is an acid.

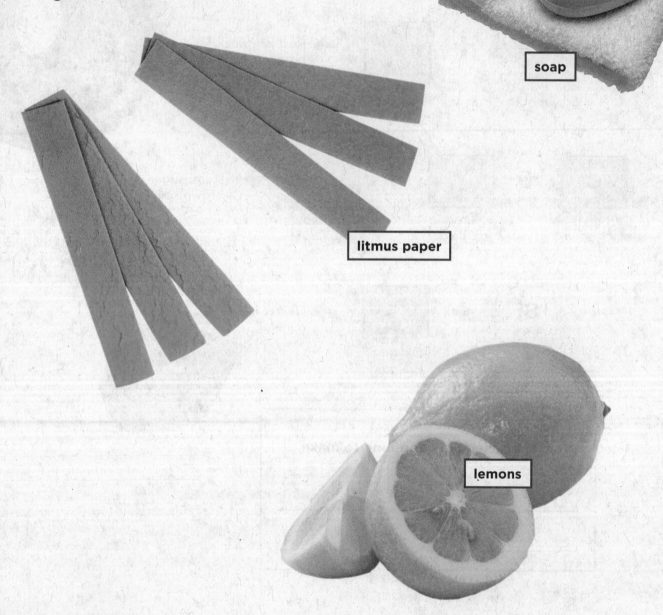

soap

litmus paper

lemons

Bases

A **base** is a substance that turns red litmus paper blue. In foods, a base will taste bitter. Drain cleaner is a strong base. It can cause harm. Soap is a base, too. You should never touch or taste something to see if it is a base.

When an acid and a base combine, they form new compounds—a salt and water. Water does not change litmus paper. Water is not an acid or a base.

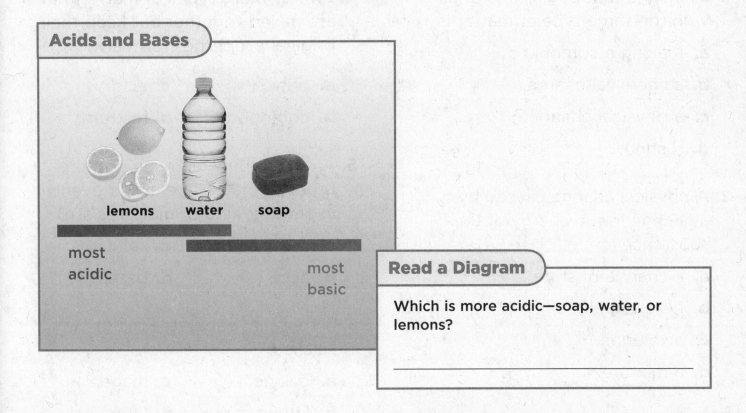

Acids and Bases

lemons   water   soap

most acidic

most basic

Read a Diagram

Which is more acidic—soap, water, or lemons?

_____

Quick Check

21. What happens to blue litmus paper when an acid touches it?

_____

22. What happens to red litmus paper when a base touches it?

_____

Matter and Its Changes

Circle the letter of the best answer.

1. A change that begins and ends with the same type of matter is

   a. making a compound.

   b. a chemical change.

   c. a physical change.

   d. rusting.

2. A physical change caused by a change in the energy of the substance is

   a. a change of state.

   b. filtration.

   c. a solution.

   d. a mixture.

3. A change that results in matter that has different properties from the original matter is a

   a. physical change.

   b. chemical change.

   c. change of state.

   d. real-world change.

4. Two or more types of matter that are mixed together but keep their original properties form a(n)

   a. base.   **c.** acid.

   b. compound.   **d.** mixture.

5. A mixture in which one or more types of matter are mixed evenly in another kind of matter is a(n)

   a. compound.   **c.** solution.

   b. acid.   **d.** base.

6. A tool that separates things by size is a

   a. balance.   **c.** magnet.

   b. litmus paper.   **d.** filter.

7. A substance that turns blue litmus paper red is a(n)

   a. base.   **c.** solution.

   b. acid.   **d.** mixture.

8. Lemons have a sour taste. They are a(n)

   a. base.   **c.** acid.

   b. solution.   **d.** mixture.

Fill in each blank with a letter to spell out the answer.

1. When solid ice melts into liquid water, the water has

    had a _c_ _h_ _a_ _n_ _g_ _e_ _o_ _f_
    _s_ _t_ _a_ _t_ _e_.

2. The change from iron to rust is a(n)

    _c_ _h_ _e_ _m_ _i_ _c_ _a_ _l_ change.

3. A salad is a(n) _m_ _i_ _x_ _t_ _u_ _r_ _e_
    of vegetables and fruits.

4. Lemonade is a(n) _s_ _o_ _l_ _u_ _t_ _i_ _o_ _n_
    of water, lemon juice, and sugar.

5. Combining sodium and chlorine forms a(n)

    _c_ _o_ _m_ _p_ _o_ _u_ _n_ _d_ called table salt.

Answer the question.

6. Is a change of state a chemical change or a physical

    change? Explain. _____

    _____

Summarize

_____

_____

_____

_____

_____

_____

CHAPTER 11

Forces

Vocabulary

 **speed** the distance traveled in an amount of time

 **force** a push or a pull

 **velocity** the speed and direction of a moving object

 **friction** a force that works against motion

 **acceleration** any change in the speed or direction of a moving object

 **gravity** a force of attraction, or pull, between objects

Why do things move?

newton a metric unit for weight, measuring an amount of force

potential energy stored energy

work the use of force to move an object

kinetic energy the energy of motion

energy the ability to do work, either to make an object move or to change matter

simple machine something that has only a few parts and makes it easier to do work

What is motion?

Think of a marble at the top of a long slanted tube. As the marble rolls down the tube, it changes location. The marble is in motion. An object is in motion if its location is changing.

Position

Position is the location of an object. You know something has moved when its position has changed.

To describe an object's position, you compare it to other objects. Words such as left, right, above, and below describe position. For example, the book is above the desk.

You can also describe an object's position by its distance from another object. Distance tells you how far apart two objects or locations are. For example, the desk's distance from the door is 3 meters.

▼ The runner is in motion. Her position is changing. Position changes as you run a distance.

Speed

A moving object has speed. **Speed** is the distance an object traveled in an amount of time.

To find an object's speed, find out how far the object moved. Then divide that distance by the time it took to go that far.

Velocity

Have you heard the term velocity? Speed and velocity are different. Speed is how fast an object is moving. **Velocity** is the speed and direction of a moving object. A train's speed might be 100 kilometers per hour. Its velocity might be 100 kilometers per hour to the east.

▲ A horse might reach a speed of 60 kilometers (40 miles) per hour.

east

▲ A train might have a velocity of 100 kilometers (60 miles) per hour to the east.

✓ Quick Check

Fill in the blanks with the correct word.

speed   direction   position

1. A change in distance over time is _____.

2. The location of an object is its _____.

3. Velocity describes an object's speed and

_____.

How do forces change motion?

A **force** is a push or a pull. Forces can make objects start or stop moving. Forces can change the speed or direction of a moving object.

Acceleration

Acceleration is any change in the speed or direction of a moving object. As ice-skaters race around a track, they go faster or slower. They turn left or right. Each time they change speed or direction, they accelerate.

Acceleration

Read a Photo

The skaters are moving around a curve in the track. Are they accelerating? Explain your answer.

_____

_____

Inertia

An ice-skater will not move unless a force acts on her. After she is moving, the skater cannot change speed or direction unless another force acts on her. Inertia (i•NUR•shuh) is the tendency of an object in motion to stay in motion or an object at rest to stay at rest.

Friction

Friction (FRIK•shuhn) is a force that works against motion. Friction acts between surfaces that touch. The surfaces rub against each other. Friction slows down or stops an object.

There is friction between a rolling ball and a table. Friction causes the ball to slow down or stop.

Friction from the table's surface is a force. It slows down the ball. ▶

✓ Quick Check

4. How does friction change the motion of a ball rolling on a table?

_____

_____

_____

What is gravity?

A force is acting on you right now. The force is gravity (GRA•vi•tee). **Gravity** is a force that acts over a distance. It pulls all objects together. Gravity's pull depends on two things:
• the amount of matter in the objects
• the distance between the objects

Objects with more mass have a stronger pull of gravity. Earth's mass is huge. Its gravity has a strong pull on all objects. The Moon has less mass than Earth. So the Moon's gravity is weaker than Earth's gravity.

Gravity is stronger when objects are closer together. As they move apart, gravity's pull is weaker.

When you jump up, Earth's gravity brings you back down to the ground. ▶

Falling to Earth

Gravity is the reason objects fall when you drop them. If you throw a ball into the air, it will not go very far. Earth's gravity will pull it down. Earth's gravity keeps things on Earth from floating off into space.

✓ Quick Check

Write true or false for each statement.

5. The pull of gravity on the Moon is stronger than the pull of gravity on Earth. _____

6. Gravity is a force that pulls all objects together. _____

7. Gravity depends only on the amount of matter in an object. _____

8. As objects move apart, the pull of gravity becomes weaker. _____

Read a Diagram

What force is pulling the apple to the ground?

_____

How do forces affect motion?

When you hit a baseball with a bat, you apply a force. Other forces may also act on the ball. These forces may be balanced or unbalanced.

Balanced Forces

A heavy book on your desk will not move. Gravity pulls the book down toward Earth. The desk pushes up on the book. The strength of the desk's push up is equal to the strength of gravity's pull down. The two forces are balanced.

Balanced forces cancel each other out when they act together on an object. Each force is equal in size and opposite in direction. Balanced forces do not cause a change in motion.

Unbalanced Forces

Let's say you want to push a heavy book across your desk. There is friction between the book and the desk. You have to overcome the friction to move the book. If you do move the book, your push is stronger than the friction.

Forces that are not equal to each other are called unbalanced forces. Unbalanced forces cause a change in motion. An object will move in the same direction as the greater force.

If the puppies pull with equal force, the ring does not move. The forces are balanced. If one puppy pulls harder, the ring moves toward him. This is an unbalanced force. ▼

Weight and Force

An object has weight because gravity's force pulls it down. Weight is also a force. Weight is measured in metric units called **newtons** (N). The newton is named for the scientist Sir Isaac Newton. More than 300 years ago, he explained the relationship between force and motion.

◄ This empty backpack weighs 5 N.

Newtons of Force

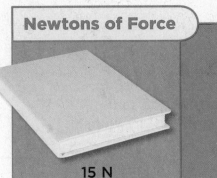

15 N

5 N

10 N

Read a Diagram

If you put the book inside the backpack, how much force would you need to lift the backpack?

_____20 N_____

Quick Check

Read each clue. Then classify the forces as balanced or unbalanced. Put a check mark in the correct column.

	Balanced Force	Unbalanced Force
9. Two puppies pull on a toy, and the toy does not move.	2	
10. Two puppies pull on a toy, and the toy moves toward one puppy.		
11. You push a heavy book across a desk.		
12. You put a heavy book on your desk, and the book does not move.		

How do forces affect acceleration?

When you run, your legs push against the ground. To run faster, you must push harder.

Forces Add Up

Look at the diagram. The red arrow shows the size of the force applied to the cart. The green arrow shows the acceleration of the load.

Look at the first two carts. When the force is increased, acceleration increases. Now look at the bottom cart. The mass is increased, but the force is not. The acceleration is less than it was before.

Read a Diagram

Why are the red arrows the same length in the first and last drawings?

_____

_____

Force and Acceleration

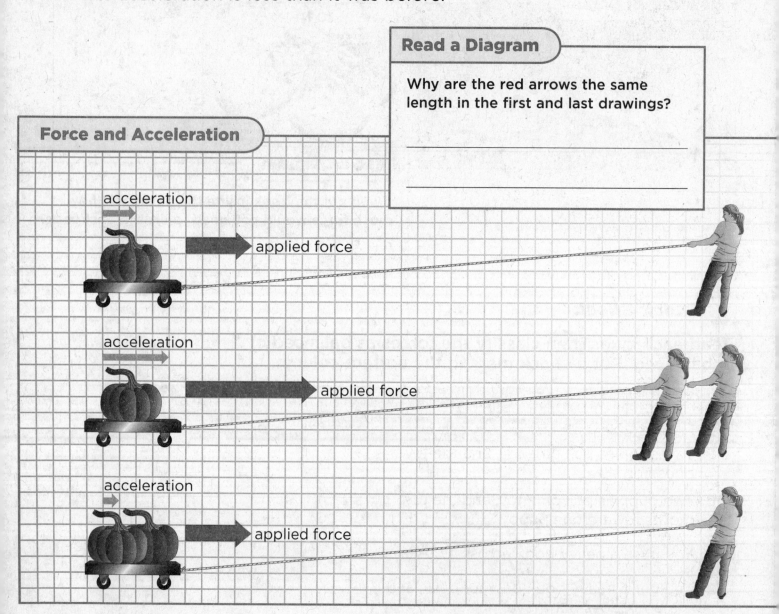

acceleration

applied force

acceleration

applied force

acceleration

applied force

LOG ON *Science in Motion* Watch accelerating masses at www.macmillanmh.com

Mass Affects Inertia

Inertia causes a moving object to stay in motion. Inertia also keeps an object that is not moving at rest. An object with more mass has more inertia. An object with less mass has less inertia.

Look at the diagram again. The load in the bottom cart has more mass than the load in the top cart. When the same force is applied, the load with greater mass accelerates half as fast. The load in the bottom cart has more inertia. It takes more force to make it move than the cart with less mass.

▲ If each racer applies the same amount of force, the racer with less mass will have greater acceleration.

✔ Quick Check

Circle the correct answer.

13. When more force is applied to the same mass, the acceleration (increases / decreases).

14. An object with more mass will have (more / less) inertia.

15. An object with less mass will have (more / less) inertia.

How does friction affect motion?

Friction works against motion. The amount of friction depends on the surfaces.

The first bobsleds had wood runners that were covered in wax. Today's bobsleds have steel runners. Steel causes less friction than wood and wax.

Ice skates have steel blades. A skater can push on a skate and glide across the ice. There is little friction between the blades and the ice.

▲ This bobsled has steel runners to reduce friction.

◄ The steel blades on the skates reduce friction on the icy surface.

◀ Friction causes the tires to push against the ground. If there were no friction, the tires would spin but the bike would not move.

If you tried to walk across the ice in sneakers, you might slide around. There is little friction between your sneakers and the ice. However, you can easily walk across a sidewalk. There is a lot of friction between your sneakers and the sidewalk. You do not slide off.

Oil can lower friction. You can add oil to the moving parts of a bike. The oil helps the parts move smoothly.

✔ Quick Check

Write true or false on the line.

16. Friction is a force that works against motion. _____

17. There is little friction between steel and ice. _____

18. Steel blades cause more friction against ice than

wood blades. _____

What is work?

When a force is used to move an object, **work** is done. Anytime you push or pull an object and the object moves, you do work.

Work can be fun! Think about a roller coaster. A roller coaster does work when it moves cars up a hill. Next, gravity does work when it pulls the cars down a hill. You do work when you walk to the roller coaster.

Force and Distance

Look at the picture. The woman is holding weights above her head. She uses a force to keep them up. However, the weights do not move. So she is not doing work.

When she lifted the weights above her head, she did work. She also does work when she lowers the weights to the ground. She must move the weights over a distance to do work.

Potential Energy

Energy (EN•uhr•jee) is the ability to do work. Stored energy is called **potential** (puh•TEN•shuhl) **energy**. Potential energy can do work in the future. The weights in the picture have potential energy. A roller coaster at the top of a hill also has potential energy. It has the ability to move down the hill in the future.

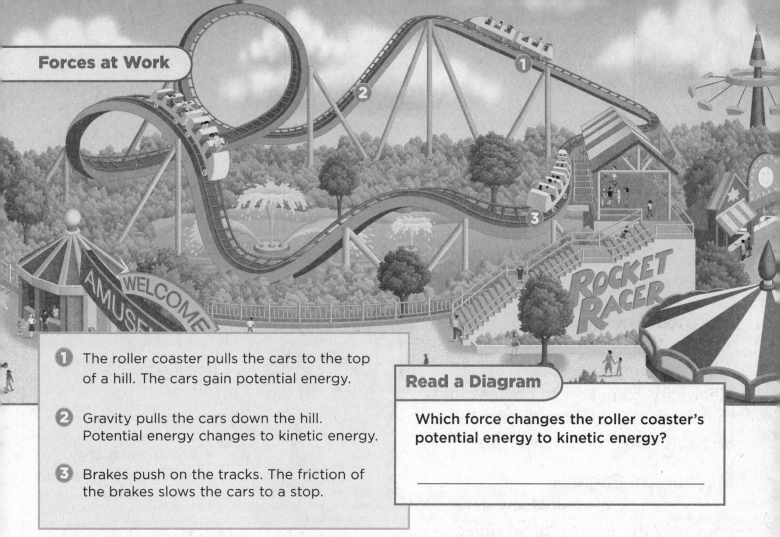

Forces at Work

1. The roller coaster pulls the cars to the top of a hill. The cars gain potential energy.

2. Gravity pulls the cars down the hill. Potential energy changes to kinetic energy.

3. Brakes push on the tracks. The friction of the brakes slows the cars to a stop.

Read a Diagram

Which force changes the roller coaster's potential energy to kinetic energy?

_____

Kinetic Energy

When an object is moving, it has energy of motion. This is called **kinetic** (ki•NET•ik) **energy.** A roller coaster has kinetic energy when it moves up or down a hill.

Potential energy can change into kinetic energy. A roller coaster that stops at the top of a hill has potential energy. When it begins to roll down the hill, the potential energy changes into kinetic energy.

✓ Quick Check

19. What is the difference between kinetic energy and

potential energy? Give an example of each. _____

_____

_____

What are some forms of energy?

Energy has many forms. What forms do you use?

Chemical Energy

Chemical energy is stored in food and fuel. When you eat food, you get the food's chemical energy. You use it to move and grow.

Electrical Energy

Electrical energy comes from moving charged particles. Most electricity comes from power plants that burn fuel. Electricity is then sent through wires to homes and schools.

Mechanical Energy

Mechanical (muh•KAN•i•kuhl) energy is the energy of an object. When an object moves, the object has kinetic energy. The movement of the object gives it mechanical energy. A roller coaster car at the top of a hill also has mechanical energy. It has potential mechanical energy.

Light Energy

The Sun, lamps, fires, and lasers are sources of light energy. Plants use light energy for photosynthesis. They use light energy to make food. Solar cells also use light energy. Solar cells change light energy into electrical energy.

◀ These grapes have chemical energy.

◀ Electrical energy makes this fan work.

These toys have mechanical energy. ▲

Nuclear Energy

Nuclear (NEW•klee•uhr) energy comes from tiny particles of matter. When the particles split apart, or join together, a lot of energy is given off.

Thermal Energy

A burning match has thermal energy. The fire gives off heat. The more thermal energy a substance has, the warmer it is.

▲ Fires have thermal energy.

Nuclear energy powers the Sun. The Sun gives off light energy. ▶

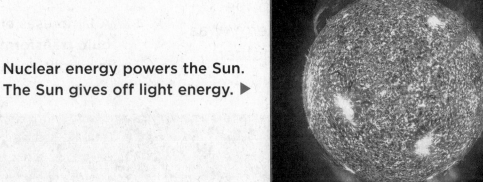

Quick Check

Draw a line from the example to the form of energy it has.

Example	Form of Energy
20. particles of matter that split apart	light energy
21. a hot oven	thermal energy
22. a windup toy	mechanical energy
23. a laser	chemical energy
24. moving charged particles	nuclear energy
25. stored food in plants	electrical energy

How can energy change?

Energy can change forms. It can also move.

Transforming Energy

Energy is transformed when it changes. A car transforms chemical energy to mechanical energy. The car moves because of the fuel. A light bulb transforms electrical energy into light and thermal energy.

Transferring Energy

Energy is transferred when it passes from one object to another. Think of one marble hitting another. The first marble stops. The second marble picks up the kinetic energy and moves. The energy was transferred from one marble to another.

▲ A lamp uses electricity. The light bulb transforms electrical energy into light and thermal energy.

A baseball bat transfers mechanical energy to the ball.

Transforming Energy

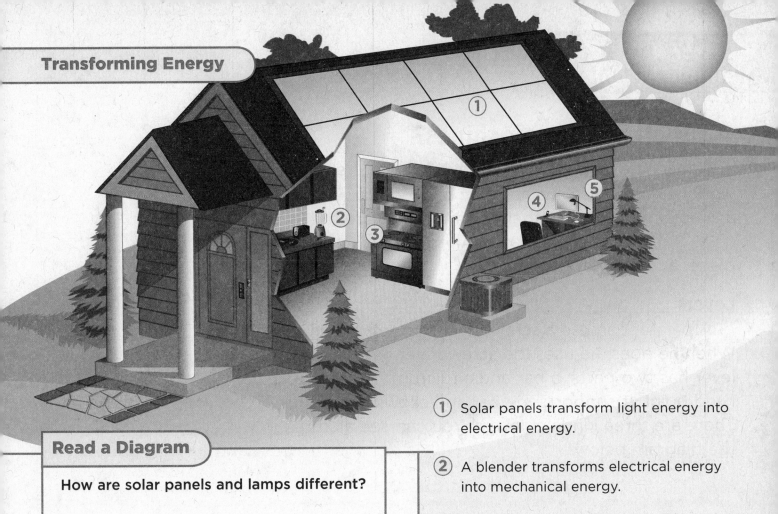

① Solar panels transform light energy into electrical energy.

② A blender transforms electrical energy into mechanical energy.

③ Electrical energy is transformed into thermal energy in a stove.

④ Electrical energy is transformed into chemical energy when a cell phone battery is charged.

⑤ A lamp transforms electrical energy into light energy.

Read a Diagram

How are solar panels and lamps different?

_____

_____

_____

✓ Quick Check

Circle the correct word to complete the sentence.

26. A light bulb (transforms / transfers) electrical energy into light energy.

27. A marble hits another marble and (transforms / transfers) kinetic energy.

What are simple machines?

A machine (muh•SHEEN) is anything that helps you do work. A **simple machine** has only a few parts. A pulley is a simple machine. Other simple machines are the lever (LEV•uhr), the wheel and axle (AK•suhl), the inclined plane, and the wedge.

Levers

A painter can use a screwdriver to open a paint can. When he does, he uses the screwdriver as a lever. A lever has two parts: a bar and a fulcrum. The fulcrum is a fixed point. It supports the bar and allows it to turn. There are three kinds of levers. You can see all three in the diagram below.

Three Kinds of Levers

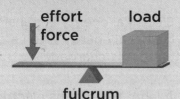

fulcrum

A *first-class lever* has its fulcrum between the load and the effort force.

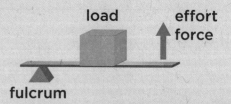

fulcrum

A *second-class lever* has its fulcrum at the end. The load is in the middle.

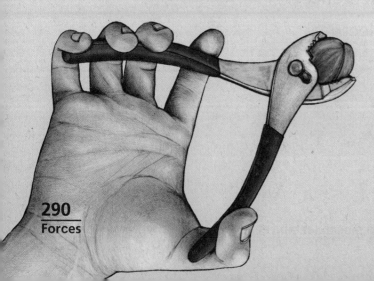

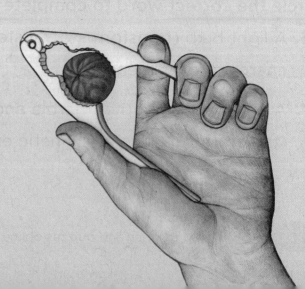

Load and Effort Force

The object being moved by a lever is called the load. The force that is used to do the work is called the effort force.

A lever does not make you stronger. It does not change how much work is done. Instead, a lever makes work easier. It changes how much effort force you use to move a load. Levers can also change the direction of a force or the distance the load is moved.

- **First-class levers** change the direction of the force. When the effort force is downward, the load moves upward.
- **Second-class levers** move the load in the same direction as the effort force. The effort force moves a long distance. The load moves a short distance.
- **Third-class levers** increase the distance the load moves. The load also moves faster and with more force. A hockey stick is a third-class lever. It hits the puck with more force than a hand alone could.

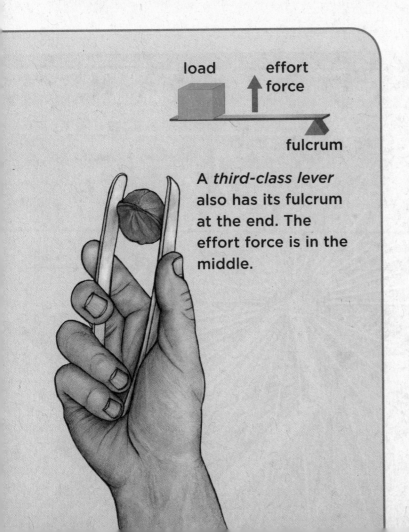

load   effort force

fulcrum

A *third-class lever* also has its fulcrum at the end. The effort force is in the middle.

✔ Quick Check

Circle true or false for each statement.

28. Simple machines do not move objects.

true   false

29. A lever has 2 parts: a fulcrum and a bar.

true   false

30. There are 3 kinds of levers.

true   false

31. A lever makes you stronger.

true   false

What are two other simple machines?

A pulley is a simple machine and so is a wheel and axle.

Wheels and Axles

A doorknob is an example of a wheel and axle. The axle is a bar that goes through the middle of the wheel.

You use a small effort force to turn the wheel. The axle turns this small force into a larger force. The larger force moves the load.

A Ferris wheel is a wheel and axle. The cars are on a giant wheel. The wheel turns on an axle. A force is applied to the axle. It moves a small distance. The axle turns the wheel a large distance.

Fixed Pulleys

A pulley makes it easier to lift a load. There are different kinds of pulleys.

The wheel of a fixed pulley is attached to something. One end of a rope goes over the wheel. The other end is tied to the load. The distance the rope is pulled is the same as the distance the load is lifted.

A fixed pulley changes the direction of a force. You pull down. The load goes up. It does not change the amount of force needed.

A Ferris wheel is a wheel and axle.

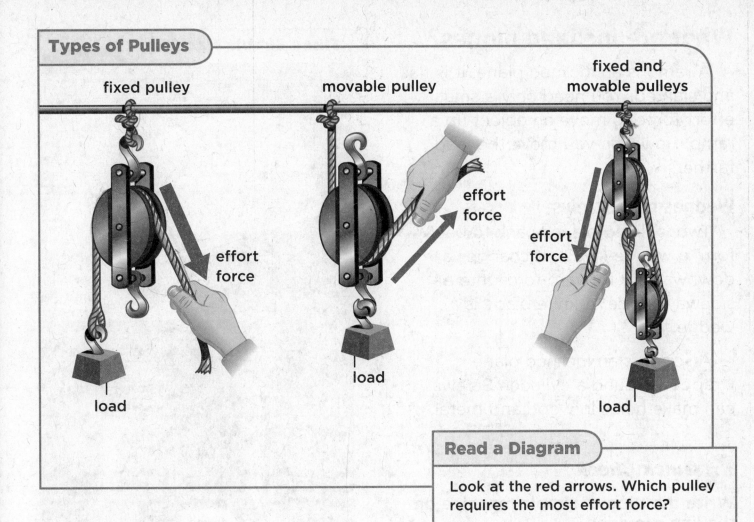

Types of Pulleys

fixed pulley

movable pulley

fixed and movable pulleys

effort force

effort force

effort force

load

load

load

Read a Diagram

Look at the red arrows. Which pulley requires the most effort force?

_____

Other Types of Pulleys

A movable pulley is attached to the load. It moves in the same direction as the load. You pull up on the rope, and the load moves up. A movable pulley uses less effort force to move a load. The distance the rope is pulled is more than the distance the load is lifted.

The third drawing shows a fixed pulley added to a movable pulley. The load moves upward. The distance it moves is less than the length of rope pulled. The effort force is downward. Less effort force is needed to lift the load.

✔ *Quick Check*

Fill in the blanks with wheel and axle, fixed, or pulley.

32. A Ferris wheel is a

_____.

33. A _____ uses a rope and a wheel.

34. A _____ pulley moves with the load.

What are inclined planes?

A ramp is an inclined plane. It is flat and slanted. You need only a small effort force to move an object up a ramp. However, you move the load farther.

Wedges and Screws

Two inclined planes back to back form a wedge (WEJ). It changes a downward or forward force into a sideways force. A knife blade is a wedge.

A screw is an inclined plane wrapped around a cylinder. Screws can make holes in wood and metal.

✔ Quick Check

Write the type of simple machine on the line provided.

35. Is a knife a screw or a wedge?

_____

36. Is a screw an inclined plane or a wedge?

_____

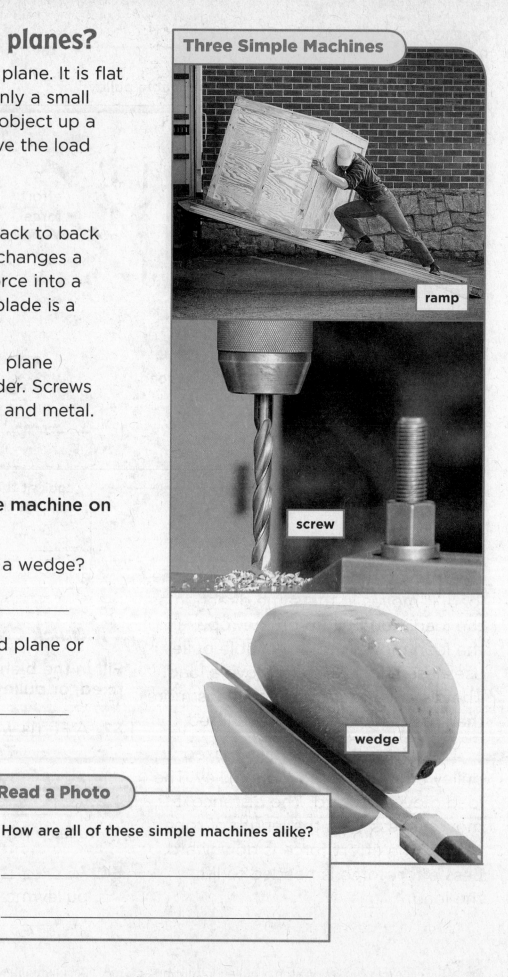

Three Simple Machines

ramp

screw

wedge

Read a Photo

How are all of these simple machines alike?

_____

_____

How do simple machines work together?

Two or more simple machines make a compound machine. Most machines are compound machines. Scissors are made of two wedges and a lever. A bike has wheels and axles. It also has levers and screws.

Efficiency

Machines do work by moving a load with less effort force. They can save time and effort. However, friction can reduce the amount of work a machine does.

Efficiency (i•FISH•uhn•see) means how well a machine is working. A machine's efficiency compares the amount of work done to the amount it should do. An efficient machine turns nearly all effort force into work.

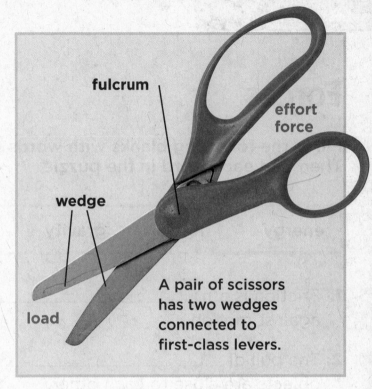

fulcrum

effort force

wedge

load

A pair of scissors has two wedges connected to first-class levers.

◀ A bike is a compound machine. The brake handles are levers. The rear tire is a wheel and axle. Screws hold parts in place.

✓ Quick Check

37. Name 3 simple machines found on a bike.

_____ _____ _____

Forces

Fill in the following blanks with words from the box.
Then find each word in the puzzle.

| energy ✓✓ | force ✓✓ | gravity ✓✓ | machine ✓✓ | speed ✓✓ | work ✓✓ |

1. Friction is a(n) _force_ that works against motion.

2. The pull of _gravity_ is stronger when objects are close to each other.

3. A roller coaster has kinetic _energy_ when it moves up a hill.

4. A lever is a simple _machine_.

5. A cheetah can reach a(n) _speed_ of 112 kilometers per hour.

6. You do _work_ when you walk.

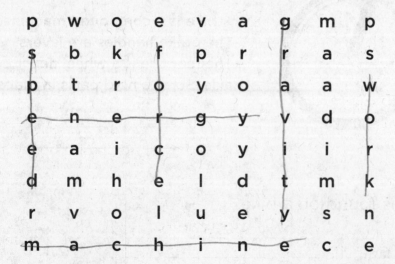

```
p  w  o  e  v  a  g  m  p
s  b  k  f  p  r  r  a  s
p  t  r  o  y  o  a  a  w
e  n  e  r  g  y  v  d  o
e  a  i  c  o  y  i  i  r
d  m  h  e  l  d  t  m  k
r  v  o  l  u  e  y  s  n
m  a  c  h  i  n  e  c  e
```

Choose the letter of the best answer.

1. The speed of an object in a certain direction is its
   a. acceleration.
   b. energy.
   c. force.
   d. velocity. ⭕

2. A force that works against motion is
   a. speed.
   b. friction. ⭕
   c. energy.
   d. gravity.

3. A metric unit for measuring an amount of force is a(n)
   a. newton. ⭕
   b. work.
   c. gravity.
   d. energy.

4. Change over time in the speed or direction of a moving object is called
   a. work.
   b. potential energy.
   c. acceleration. ⭕
   d. kinetic energy.

5. Stored energy is called
   a. potential energy. ⭕
   b. kinetic energy.
   c. work energy.
   d. simple machine.

6. The energy of motion is called
   a. potential energy.
   b. speed.
   c. force.
   d. kinetic energy. ⭕

Summarize

_____

_____

_____

_____

CHAPTER 12

Energy

Vocabulary

 **heat** the flow of thermal energy from warmer to cooler objects

 **conduction** the transfer of thermal energy between materials that are touching

 **convection** the transfer of thermal energy in gases or liquids

 **radiation** the transfer of thermal energy through space

 **sound wave** a wave that travels along or through matter, produced by a vibration

 **echo** a sound wave that has bounced off a surface

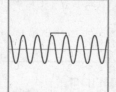 **wavelength** the distance from the top of one wave to the top of the next

 **frequency** the number of wavelengths that pass a point in a given amount of time

 **pitch** the highness or lowness of a sound

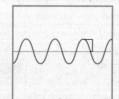 **amplitude** the height of a wave

How do we use energy?

refraction the bending of light as it passes from one material to another

reflection the bouncing of light or sound waves off a surface

transparent letting all the light through so that objects on the other side can be seen clearly

translucent letting only some light through so that objects on the other side appear blurry

opaque completely blocking light from passing through

circuit a path along which electric current flows

current electricity the flow of electrical charges through a circuit

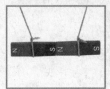

pole one of two ends of a magnet, where a magnet's pull is strongest

What is heat?

Thermal energy makes the particles in matter move. Heat is the flow of thermal energy. **Heat** always moves from warmer objects to cooler ones.

Transferring Heat

Hot particles move quickly. They bump into cooler particles. This transfers thermal energy. The warm particles slow down after they transfer thermal energy. The cooler particles speed up. The temperature gets warmer. In time, all the particles move at the same speed.

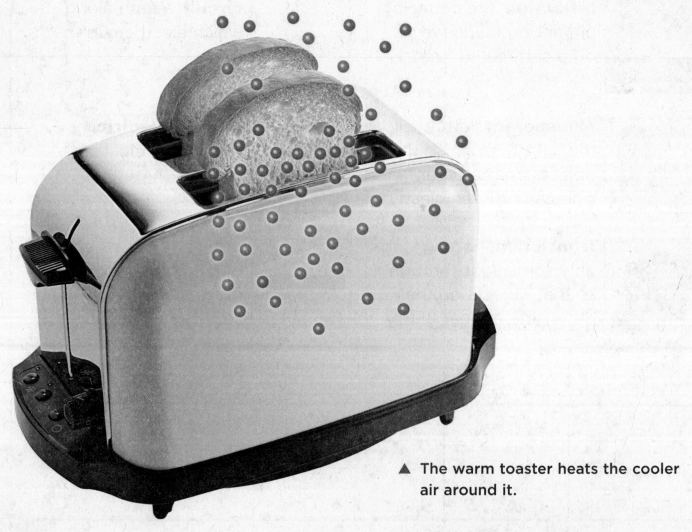

▲ The warm toaster heats the cooler air around it.

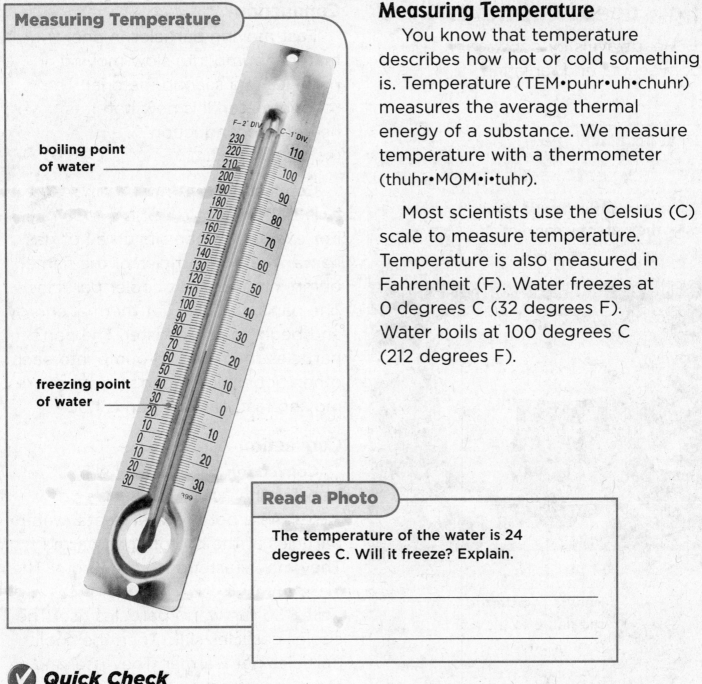

boiling point
of water

freezing point
of water

Measuring Temperature

You know that temperature describes how hot or cold something is. Temperature (TEM•puhr•uh•chuhr) measures the average thermal energy of a substance. We measure temperature with a thermometer (thuhr•MOM•i•tuhr).

Most scientists use the Celsius (C) scale to measure temperature. Temperature is also measured in Fahrenheit (F). Water freezes at 0 degrees C (32 degrees F). Water boils at 100 degrees C (212 degrees F).

Read a Photo

The temperature of the water is 24 degrees C. Will it freeze? Explain.

_____

_____

✔ Quick Check

Write the effect for each cause.

1. Cause: Warm particles transfer thermal energy to cooler particles.

Effect: Warm particles _____.

2. Cause: Cool particles bump into warmer particles.

Effect: Cool particles _____.

How does heat travel?

Heat travels by conduction, convection, and radiation.

Conduction

Fast-moving particles in one material bump into slow-moving particles in a second material. The second material is now hot! It is heated by conduction (kuhn•DUK•shuhn).

Conduction transfers thermal energy between materials that are touching. For example, a pan sits on a hot gas flame. The hot particles of the flame bump into the pan's cooler particles. The pan's particles get thermal energy and begin to move faster. The pan's particles then start to bump into each other. Soon, all the pan's particles are moving fast, and the pan is hot.

Convection

Convection (kuhn•VEK•shuhn) transfers heat through liquids or gases. As a pot of water heats, water particles at the bottom get warm first. They move faster and spread out. The warm water is less dense than the cool water. So the warm particles rise. The cooler particles sink. Then the cooler particles get warmer. They rise, and the cycle repeats.

Heat Transfer

Heat is transferred through the water by convection.

Heat is transferred from the flame to the pot by conduction.

Read a Diagram

In the pot of water, do the warmer particles sink to the bottom or rise to the top?

_____

Radiation

Radiation (ray•dee•AY•shuhn) transfers thermal energy through space. Without radiation, the Sun's energy would not reach Earth.

Insulators and Conductors

Different materials transfer heat at different rates. Insulators transfer heat slowly. Fat is an insulator. Conductors transfer heat quickly. Metal is a good conductor.

A copper kettle is a good conductor for hot liquids. ▼

◄ Wool mittens are good insulators for your hands.

✔ Quick Check

Draw a line from the definition to the word.

3. the transfer of heat between materials that are touching

5 radiation

4. the transfer of heat through a liquid or a gas

3 conduction

5. the transfer of heat through space

4 convection

How does heat change matter?

Heat causes physical changes. Heat also causes chemical changes.

Expanding and Shrinking

When you heat matter, its particles usually move farther apart. The matter expands and takes up more space. When you cool matter, its particles usually move closer together. The matter shrinks and takes up less space. Expanding and shrinking are physical changes.

Chemical Changes

Heat can cause some matter to burn. Burning is a chemical change. Energy is released when matter burns.

◀ Heat can cause a match to burn. The match is going through a chemical change.

Changes of State

If you add enough heat, matter can change state. Changes of state are physical changes. In the picture below, solid metal melts into a liquid. If more energy were added, the liquid metal would change to a gas.

◀ Heat can change solid metal to a liquid.

✔ Quick Check

Fill in the blanks to complete each cause-and-effect sentence. Use words from the word bank.

burn ✓	expands ✓	faster ✓	heat ✓	shrinks ✓

6. When you add heat to matter, the particles move

_____faster_____ and farther apart.

7. Matter _____expands_____ and takes up more

space when energy is added.

8. When it is cooled, most matter _____shrinks_____,

or gets smaller.

9. Heat can cause a match to _____burn_____.

10. Solid metal can melt into a liquid when

_____heat_____ is added.

LOG ON ⓔ**-Review** Summaries and quizzes online at www.macmillanmh.com

What is sound?

When you pluck a guitar string, it moves back and forth. This motion is called a vibration (vye•BRAY•shuhn). All sounds begin with a vibration. The vibrating guitar string pushes on nearby air particles. The sound travels outward. The air particles begin to move.

When the moving air particles reach a person's ear, a sound is heard. Sound can travel through air, water, metals, and glass. Sound can travel through any kind of matter.

When the drummer strikes the drum, it vibrates. The vibration makes a sound wave. ▼

Sound Waves

Look at the alarm clock picture. The picture shows a model of sound waves. Some air particles are crowded close together. Some air particles are spaced far apart. The air particles move as sound waves. **Sound waves** result from vibrations. Sound waves travel along or through matter.

Sound waves spread outward from the bell. The sound waves spread outward in all directions.

Energy Transfer

To make an object vibrate, you need energy. Sound waves carry that energy away from the vibrating object. The waves transfer the energy from particle to particle.

As the waves travel outward, they lose energy. The closer a person is to the vibration, the louder the sound.

◄ A ringing bell sends sound waves in all directions.

✔ Quick Check

11. All sound begins with a _____.

12. A wave that travels along or through matter

    is called a _____.

How does sound travel?

You know that sound travels through the air. Sound can move through solids, liquids, and gases.

Echoes

Sometimes sound waves bounce off a surface. The surface reflects the sound. This makes the sound come back. An **echo** (EK•oh) is a reflected sound. If the reflecting surface is far away, the echo takes longer to bounce back.

Dolphins use echoes to find prey. Fish and other objects reflect sounds that the dolphins make.

Dolphins use echoes to find fish to eat.

Sound Speeds

Read a Diagram

Does sound travel faster in iron or in rubber?
_____

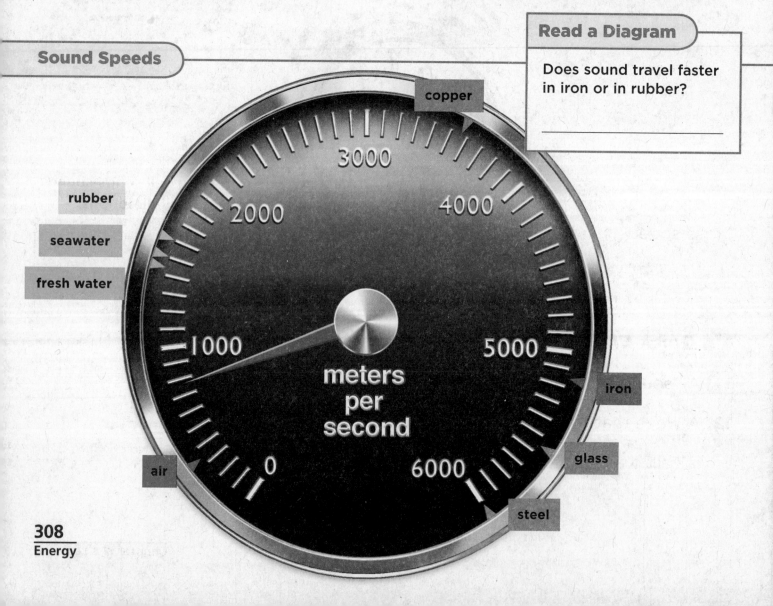

copper

rubber

seawater

fresh water

3000

2000

4000

1000

5000

meters per second

0

6000

iron

glass

air

steel

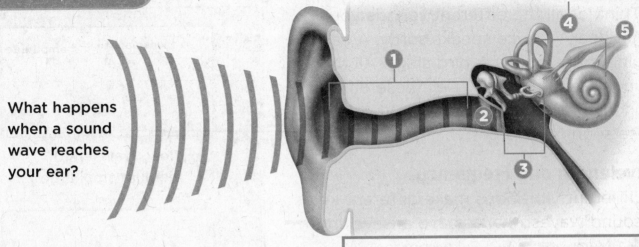

What happens when a sound wave reaches your ear?

1. **outer ear** The outer ear collects sound waves.

2. **eardrum** Sound waves make the eardrum vibrate.

3. **middle ear** The vibrations are picked up by three tiny bones in the middle ear.

4. **inner ear** A coiled tube in the inner ear receives the vibrations. The tube is filled with fluid and is lined with tiny hair cells.

5. **nerve to brain** The hair cells signal a nerve in the ear. The nerve carries signals to the brain. The brain interprets the signals as sound.

The Human Ear

Sound waves carry sound energy to tiny organs in the ear. The energy makes these organs vibrate. The diagram shows how the ear turns vibrations into what we hear as sound.

The Speed of Sound

Sound does not travel at the same speed through all materials. It travels slowest in a gas, such as air. Sound travels fastest through a solid. The diagram to the left shows some of these speeds.

 **Quick Check**

13. How do dolphins use echoes?

_____

_____

How do sounds differ?

Think of all the different sounds you might hear. A voice speaks softly. A car engine roars loudly. A bird sings. All the sounds are different. Yet all these different sounds come from vibrations. Why are these sounds different?

Wavelength and Frequency

Different vibrations make different kinds of sound waves. Waves have a wavelength and a frequency (FREE•kwuhn•see). A **wavelength** is the distance from the top of one wave to the top of the next wave. In a sound wave, this is the distance from one area of crowded particles to the next.

Frequency is the number of vibrations something makes in a given amount of time. When you strike a small bell, it vibrates quickly. The vibrations produce sound waves with a high frequency. A bass guitar makes sound waves with a low frequency. The bass guitar strings vibrate slowly.

✔ Quick Check

14. The distance from the top of one wave to the top of the next wave is

a wave's _____.

15. The number of vibrations a sound source makes in a given amount of

time is its _____.

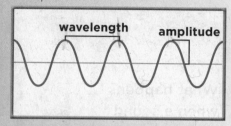

long wavelength
medium amplitude

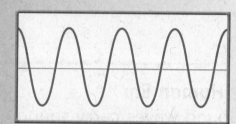

long wavelength
high amplitude

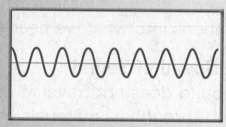

short wavelength
low amplitude

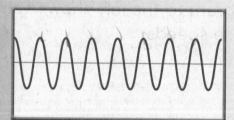

short wavelength
medium amplitude

Read a Diagram

Which sound wave will have the highest volume?

_____

Pitch

The frequency of a sound wave determines its pitch. **Pitch** means the highness or lowness of a sound. High sounds, like the sound of a mosquito, have a high frequency. Low sounds, like the croaks of a frog, have a low frequency.

The pitch of a guitar can change when you change the strings. Shorter, tighter, or thinner strings vibrate more quickly. The sounds have a higher pitch.

Amplitude and Volume

Amplitude (AM•pli•tewd) is a wave's height. Amplitude affects volume, or the loudness, of a sound. Sound waves with high amplitude will have a high volume. Sounds with low amplitude will have a low volume.

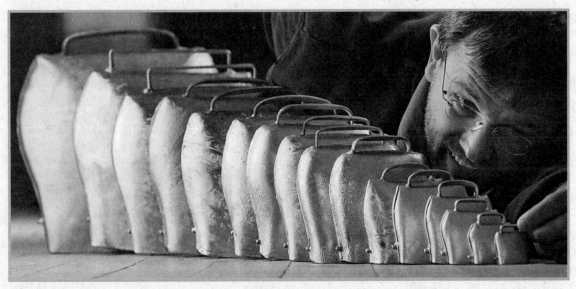

▲ The size of each bell determines the pitch of the sound.

✓ Quick Check

Read each sentence. Circle true or false.

16. Sound travels fastest through air.   true   false

17. Sound cannot travel through water.   true   false

18. A sound wave's frequency determines   true   false
its pitch.

What is light?

Light is a form of energy that lets us see objects. Light comes from the Sun, light bulbs, fire, and other sources. All the colors you see are part of light.

The Visible Spectrum

White light is made of colored light. Look at the prism (PRIZ•uhm). A prism is an object that separates white light into bands of colored light. These colors make up the visible (VIZ•uh•buhl) spectrum.

The visible spectrum is not the only part of light. Like sound, light travels in waves. The electromagnetic spectrum is the range of waves that make up light.

Wavelengths and Energy

The light waves in the electromagnetic spectrum have different wavelengths. Each wavelength carries a different amount of energy. The longer the wavelength, the less energy the wave has.

Light waves can be both helpful and harmful. Ultraviolet, or UV, waves are dangerous. They can burn your skin. X-ray waves help doctors see inside the body.

Where are visible light waves in the electromagnetic spectrum? Look for them in the diagram.

The Electromagnetic Spectrum

radio waves   microwaves   infrared waves

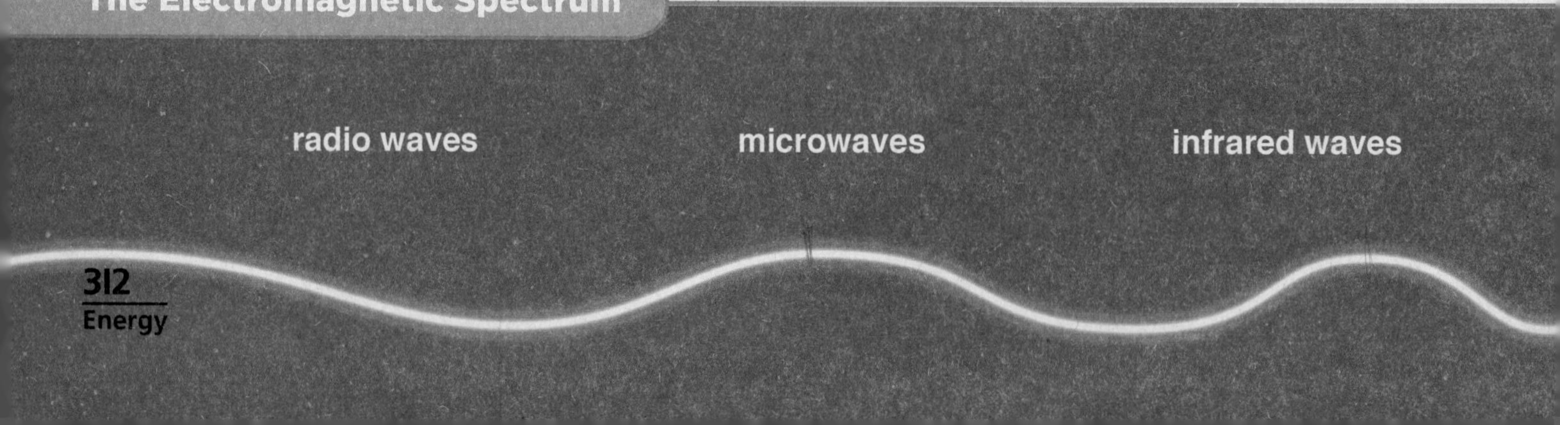

Wavelengths of Light

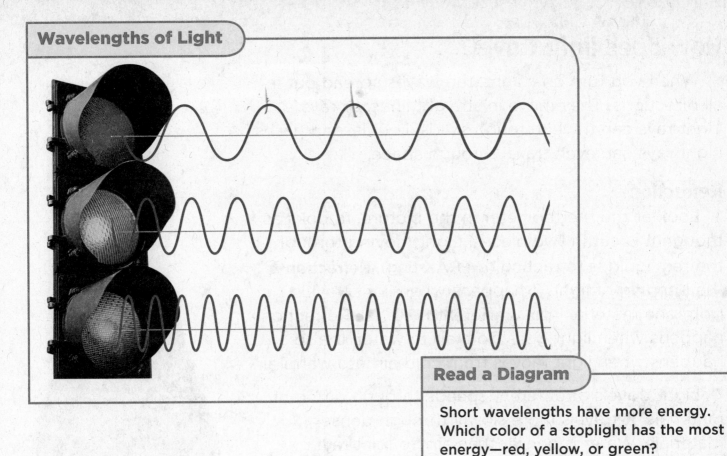

Read a Diagram

Short wavelengths have more energy. Which color of a stoplight has the most energy—red, yellow, or green?

_____

✓ Quick Check

19. Along the electromagnetic spectrum, _____ waves have the shortest wavelength.

20. The colors we can see make up the _____ .

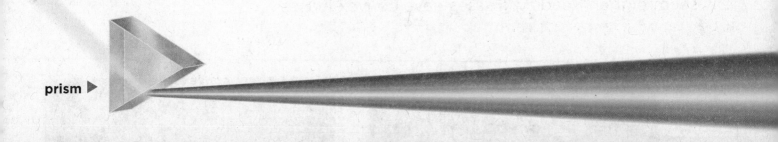

prism ▶

visible waves   ultraviolet waves   x-ray waves   gamma waves

How does light travel?

When you turn on a light, the waves spread out in all directions. They move in straight lines, or rays. Light rays can travel through solids, liquids, and gases. Light rays can even travel through space.

Refraction

Look at the thermometer in the picture. It looks as though it is cut in two pieces. Light moving through the red liquid is refracting (ri•FRAK•ting). **Refraction** is the bending of light. It happens when light travels from one material into a different material. Refraction happens when light goes from air to water. It also happens when light moves from cold air into warm air.

Light travels at different speeds through different materials. It travels more slowly through denser materials. Water is denser than air. So light rays refract, or bend, where water and air meet.

A lens is a tool that refracts light. A concave lens curves inward. Light rays bend outward from the center of the lens. The rays spread apart. Glasses that help you see faraway objects are made with concave lenses.

A convex lens bulges outward. Light rays bend inward toward its center. This makes an object near the lens seem bigger. Reading glasses have convex lenses.

▲ Refraction makes the thermometer appear to be in two pieces.

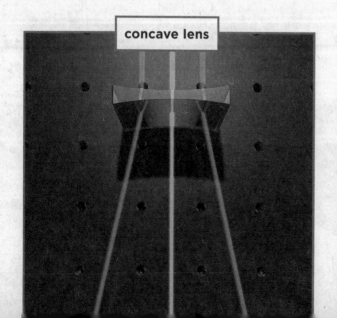

concave lens

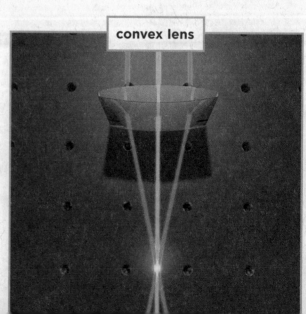

convex lens

The Human Eye

Light bounces off objects. When this light enters your eye, you see the object.

To see an object, light must first pass through a thin, clear tissue called the cornea. Next, light passes through the hole in the center of the eye, called the pupil (PYEW•puhl).

The iris (EYE•ris) is the colored part of the eye. The iris controls the amount of light that enters the pupil.

From the pupil, light moves through the lens. The lens refracts the light and focuses the image onto the back of the eye. The image is upside down.

The back of the eye is called the retina (RET•uh•nuh). The retina changes the upside-down image into signals. A nerve sends the signals to the brain. The brain interprets the signals as right-side up.

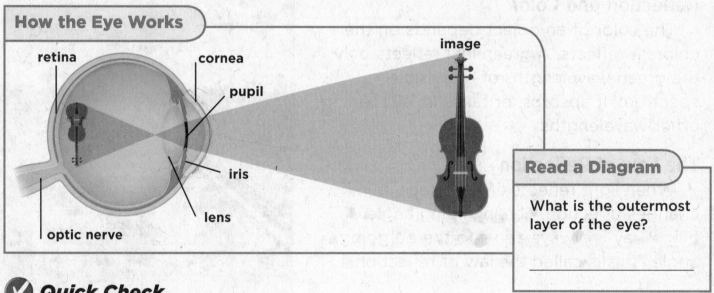

How the Eye Works

retina

cornea

pupil

iris

lens

optic nerve

image

Read a Diagram

What is the outermost layer of the eye?

_____

✔ Quick Check

Correct these false statements.

21. Glasses that help you see far away are made with convex lenses.

_____

22. Reading glasses make objects seem bigger because they are made with concave lenses.

_____

23. Refraction happens when light rays are straightened.

_____

What is reflection?

Like sound waves, light waves can bounce off a surface. **Reflection** is a wave that hits a surface and bounces off. Most of the light that reaches your eyes is reflected light.

Surfaces That Reflect Light

Most surfaces reflect at least some light. Smooth, shiny surfaces reflect almost all the light that hits them. Dull, rough surfaces reflect less light.

Reflection and Color

The color of an object depends on the colors it reflects. A green leaf reflects only the green wavelengths of the visible spectrum. It absorbs, or takes in, all the other wavelengths.

The Law of Reflection

When light reflects off a surface, it changes direction. The incoming angle of a light ray is always equal to the outgoing angle. This is called the law of reflection.

The river is smooth enough to reflect light.

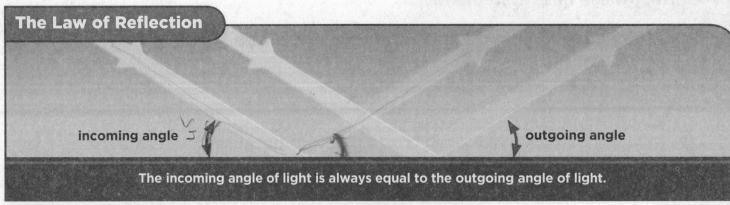

The Law of Reflection

incoming angle   outgoing angle

The incoming angle of light is always equal to the outgoing angle of light.

 **Quick Check**

24. What light color does a green leaf reflect? _____

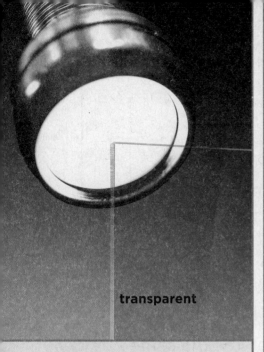
transparent

translucent

opaque

Glass lets light pass through it.	Plastic scatters light in different directions.	Wood prevents light from getting through.

What can light pass through?

When light strikes an object, it may or may not pass through.

Transparent Objects

Some materials are **transparent.** They allow light to pass through. Glass and water are transparent. You can see clearly through them.

Translucent Objects

Translucent (trans•LEW•suhnt) materials scatter light in different directions. It is hard to see clearly through translucent materials. Cloudy, plastic shower doors are translucent.

Opaque Objects

Opaque (oh•PAYK) materials block light completely. Wood and metal are opaque.

✔ *Quick Check*

Match the correct material property with its description.

25. lets light pass through it   opaque

26. scatters light in different directions   transparent

27. prevents light from getting through   translucent

What is electrical charge?

Electricity is a moving electrical charge. An electrical charge is not something you can see, smell, or weigh. It is a property of matter.

Positive and Negative Particles

Matter is made up of atoms. Atoms are made of even smaller particles. Some have a positive electrical charge (+). Others have a negative electrical charge (-).

Charges Interact

A positive charge and a negative charge attract, or pull, each other. Like charges repel, or push away from, each other. Positive repels positive. Negative repels negative.

Most matter has the same number of positive and negative charges. The charges cancel each other out. This means that the matter is neutral (NEW•truhl).

Overall Charge

③ The negative charges on the balloon attract the positive charges on the wall. The balloon sticks to the wall.

② By rubbing the balloon with the wool, negative charges build up on the balloon.

① A balloon and a wool cloth are neutral. Each has as many negative charges as positive charges.

Charges Add Up

When two objects touch, charged particles can move from one object to the other. Negative charges move more easily than positive ones.

If you rub a balloon with a wool cloth, negative charges move from the wool to the balloon. The balloon gets a buildup of negative charges. A buildup means an object has more of one kind of charge than another object. The wool gets a buildup of positive charges.

④ In time, the charges move around. The balloon becomes neutral. It is no longer attracted to the wall, so it falls.

Static Electricity

The buildup of electrical charges is called static electricity. Rubbing objects together causes static electricity. A balloon with a negative charge will attract the positive charges in a wall. So the balloon will stick to the wall. Over time, the charges move around. The balloon becomes neutral and falls.

✓ Quick Check

Look at the picture of balloons 1, 2, 3, and 4. Then answer the questions.

28. What causes balloon 2 to have a buildup of negative charges?

_____

29. Does balloon 3 have a positive or a negative charge?

_____

30. Does balloon 4 have a charge or is it neutral?

_____

31. Which balloons have static electricity?

_____

How do charges move?

If you walk across carpet and then touch a metal doorknob, you may feel a shock. It is caused by a fast movement of charged particles.

Electrical Discharge

When you move across the carpet, negative charges rub off the carpet onto you. Your body gets a buildup of negative charges. They build up until you touch something. Then they move to the object you touch. This sudden movement is called discharge (DIS•charj).

Lightning

Lightning is the discharge of static electricity during a storm. Inside a storm cloud, ice and water droplets rub against each other. Positive and negative charges build up in different places. If the buildup gets large enough, charges jump as lightning.

Lightning is the discharge of static electricity.

Electric Current

Electrical charges can flow like water. A flow of electrical charges is electric current.

Circuits

Electric current needs a path. The path along which electric current flows is called a **circuit** (SIR•kit). A simple circuit has three parts:

• a power source, such as a battery
• a load, such as a light bulb
• connectors, such as wires

The flow of electrical charges through a circuit is called **current electricity.** An unbroken circuit is a closed circuit. Current flows through a closed circuit. A circuit with gaps or breaks is an open circuit. Current will not flow through an open circuit.

Switches

Many circuits have a switch. A switch turns current electricity on and off. Lights in your classroom may have a switch. When it is on, the circuit is closed. Current flows through. The lights shine. When the switch is off, the circuit is open. No current flows. The lights are off.

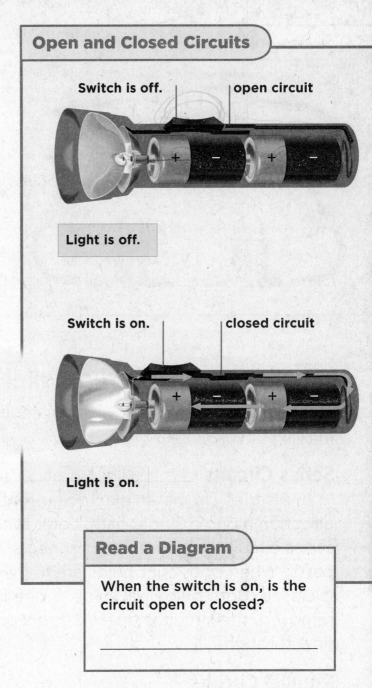

Open and Closed Circuits

Switch is off.   open circuit

Light is off.

Switch is on.   closed circuit

Light is on.

Read a Diagram

When the switch is on, is the circuit open or closed?

_____

✔ *Quick Check*

32. The path along which an electric current flows is a(n)

_____.

33. The flow of electrical charges through a circuit is

_____.

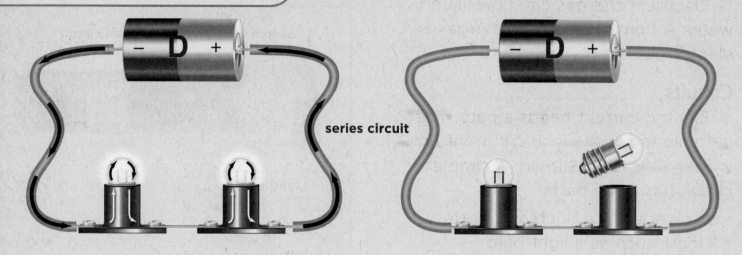

series circuit

What are series and parallel circuits?

There are two kinds of circuits: series circuits and parallel circuits.

Series Circuits

In a series circuit, an electric current flows in the same direction along a single path. Look at the diagram of the series circuit. One wire loop connects all the circuit parts. When both light bulbs are in place, the circuit is closed. Both bulbs will shine. If one light bulb is removed, the circuit is open. Current will not flow through.

Parallel Circuits

In a parallel circuit, electric current flows through more than one path. These paths are often called branches. They divide the electric current.

Look at the diagram of the parallel circuit. Both light bulbs connect to the power source. There are two different paths. If one light bulb is removed, the other bulb still lights. Current still flows through the complete circuit.

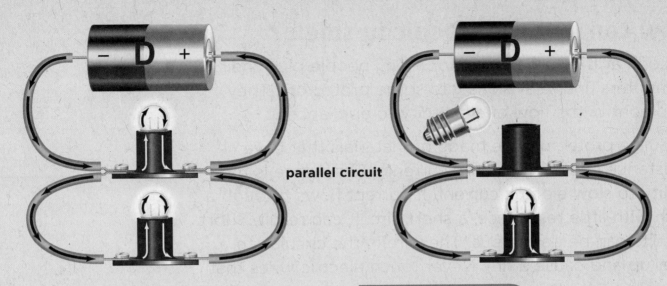

parallel circuit

Read a Diagram

Which circuit has more than one path?

_____

Circuits in Homes

Electrical outlets in most homes are connected in parallel circuits. When you turn off one light, other lights and appliances stay on. If outlets were connected by a series circuit, all the electricity would turn off at once.

✓ Quick Check

34. Why is a parallel circuit better to use in a home?

_____

_____

_____

LOG ON *Science in Motion* Learn more at www.macmillanmh.com

How can you use electricity safely?

Look at the surge protector. Many people plug their computers and televisions into surge protectors. They help control the flow of electricity to appliances.

Surge protectors are made of materials that have resistance to strong electric currents. Resistance is the ability to slow electric current. If current flows through a path with little resistance, a short circuit can result. Short circuits can be dangerous. The wire in the circuit can heat up and cause a fire. Never touch electric wires that are torn.

▲ A surge protector protects electrical devices from too much electricity.

Fuses and Circuit Breakers

A fuse is a device that helps prevent short circuits. A fuse has a thin strip of metal. The metal has high resistance. If too much current flows through, the strip heats up and melts. The circuit opens. This stops current from flowing.

Fuses can be used only once. But circuit breakers can be reset. A circuit breaker is a switch that protects circuits. When a high current flows through a circuit breaker, the switch opens. The current stops flowing.

Homes can have fuses or circuit breakers. Do you know which you have in your home?

If a fuse breaks, it cannot be used again.

Most homes have circuit breakers.

✓ Quick Check

Circle true or false for each sentence.

35. A fuse can be used over and over again after it breaks.　　true　　false

36. A circuit breaker can be reset.　　true　　false

37. Short circuits can cause fires.　　true　　false

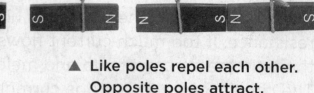

What is a magnet?

A magnet is an object that has a magnetic field. It can attract some metals.

Two magnets can pull, or attract, each other. Two magnets can also push away, or repel, each other. This push and pull of a magnet is a magnetic force.

Magnetic Poles

The force of a magnet is strongest at each **pole.** All magnets have a north pole and a south pole. Poles are labeled N and S. A magnet's north pole attracts the south pole of another magnet. Like poles repel each other. The attraction of two magnets is strongest when the magnets are close together. Magnetic force gets weaker with distance.

Magnetic Particles

Some metals are made of tiny particles that act like small magnets. If an object made of iron is close to a magnet, the iron particles line up. The object becomes a temporary magnet.

▲ Like poles repel each other. Opposite poles attract.

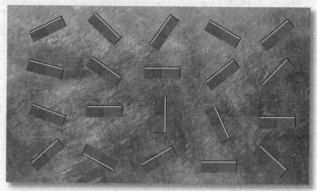

▲ Metals are made of tiny particles. Normally, the particles point in different directions.

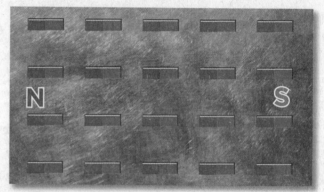

▲ When a magnet is brought near iron, nickel, or cobalt, the particles line up. They point in the same direction.

✓ Quick Check

Look at the bar magnets and their poles. Write repel or attract on the lines.

38.  _____

39. | N | S | | S | N | _____

What are magnetic fields?

A magnetic field is the area of magnetic force around a magnet. Earth is a giant magnet. Much of Earth's core is made up of melted iron. This iron makes a magnetic field around Earth.

Earth has two North Poles. The geographic North Pole is at one end of Earth's axis. The magnetic north pole is near the geographic North Pole. However, the two locations are not the same. This is also true of the South Pole.

Using a Compass

Earth's magnetic north pole attracts the south pole of a compass needle. This is why a compass needle always points north. You can make a compass with a bar magnet and a piece of string. Hang the magnet from the string. The magnet will line up with Earth's magnetic field.

▼ The lines show Earth's magnetic field. They are called magnetic field lines.

Earth's Magnetic Field

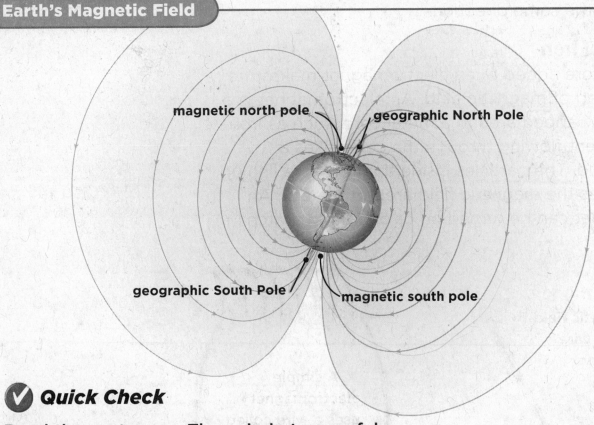

magnetic north pole

geographic North Pole

geographic South Pole

magnetic south pole

✔ Quick Check

Read the sentences. Then circle true or false.

40. The inside of Earth is melted cobalt.　　true　　false

41. The area of magnetic force around a magnet is a magnetic field.　　true　　false

42. Earth is surrounded by a magnetic field.　　true　　false

What is an electromagnet?

Electric current is a flow of charged particles. When charged particles move, they form magnetic fields. You can use electric current to make a magnet.

The Effect of Current

Electric current moving through a wire sets up a magnetic field around the wire. When the current is turned off, the magnetic field goes away.

The Effect of Coils

A wire can be wrapped into a long coil. Electric current flows through the wire. The magnetic field around the coil is stronger than a straight wire. Each loop in the coil is like a little magnet. The loops all pull and push in the same direction.

The Effect of Iron

A metal core added to an electromagnet makes the strongest kind of magnetic field. An electromagnet is a coil of wire wrapped around a metal core, such as iron. Electric current flowing through the coil makes a magnetic field. The particles inside the iron core line up. This increases the magnetic field around the coil. An electromagnet can be turned on or off with a switch.

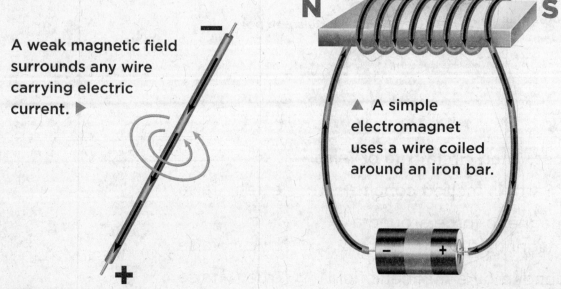

A weak magnetic field surrounds any wire carrying electric current. ▶

▲ A simple electromagnet uses a wire coiled around an iron bar.

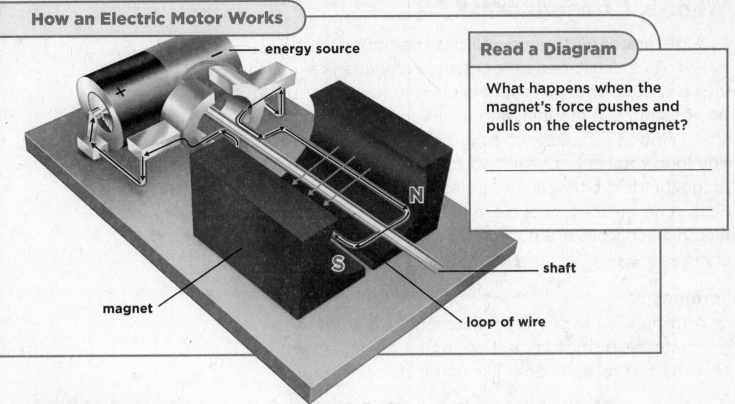

energy source

N

S

magnet

shaft

loop of wire

What happens when the magnet's force pushes and pulls on the electromagnet?

_____

_____

Electric Motors

Electromagnets are used to power electric motors. An electric motor is a device that changes electrical energy into mechanical energy, or motion. A simple electric motor has three parts. It has a power source, a magnet, and a wire loop. The wire is attached to a rod called a shaft. The shaft can spin.

The power source makes electric current. The current runs through the wire loop. The loop becomes an electromagnet. The magnet pushes and pulls on the electromagnet. The magnetic force causes the loop and shaft to spin. The spinning shaft can then attach to a wheel or gear.

✔ Quick Check

Order the events for making an electromagnet. Write 1, 2, or 3 next to each sentence.

43. ____ Current runs through the coil, which sets up a magnetic field.

44. ____ Turn on the power source to make an electric current.

45. ____ Wrap a wire around a metal core, and attach the wire to a power source.

What is a generator?

A generator (JEN•uh•ray•tuhr) is the opposite of a motor. An electric generator changes mechanical energy into electrical energy. A simple electric generator has a power source, a magnet, and a wire loop attached to a shaft. Motion is needed to turn the shaft and wire loop. The loop rotates between two magnetic poles. The magnetic field between the poles makes electric current. Each time the loop gets close to the magnet's poles, electrical charges are pushed through. These moving charges are electric current.

Turbines

A turbine is the source of mechanical energy for a generator. A turbine is a set of angled blades attached to a shaft. A turbine looks like an electric fan.

Steam, water, or air is used to turn the blades. The turning blades spin the shaft. It then spins the wire loop or magnet inside the generator.

How A Generator Works

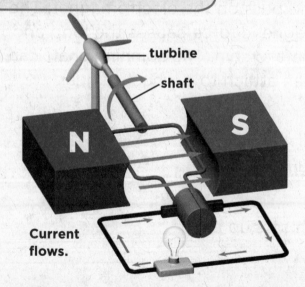

Mechanical energy turns the blades of the turbine. The blades turn the shaft. The shaft spins the wire loop. The loop spins between the magnetic poles.

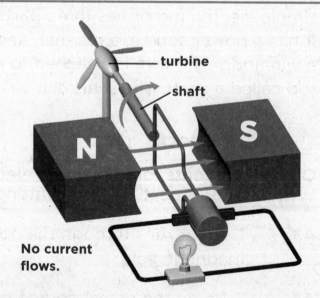

As the wire loop spins, it moves outside of the magnetic field. The circuit is open for less than one second. The loop turns quickly. You cannot see the light flicker off and on.

Alternating Current

We depend on generators to produce nearly all of our electrical energy. Most electric generators make an alternating current, or AC. With AC, electrical charges continuously flow in one direction and then flow in the opposite direction. Most electrical wall outlets in your home and school use AC.

Direct Current

When electric current flows in only one direction, it is called direct current, or DC. Like AC, direct currents flow continuously. However, the charges do not stop or reverse direction. A battery is an example of a DC power source.

 **Quick Check**

46. How is a generator the opposite of a motor?

_____

_____

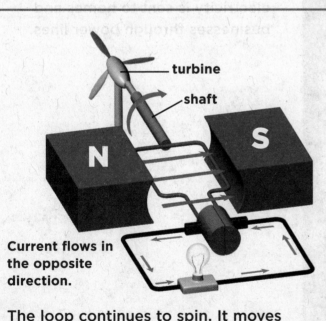

turbine

shaft

N   S

Current flows in the opposite direction.

The loop continues to spin. It moves into the magnetic field again. The poles of the loop face opposite magnets. The current reverses direction.

Read a Diagram

What turns the shaft and produces the mechanical energy for the electric generator?

_____

How does electricity get to your home?

Power plants make electrical energy. Electric current carries the energy to homes. The current moves in a circuit. The circuit connects to wall outlets.

Voltage

Voltage (VOHL•tij) is the strength of a power source. It is measured in volts. To prevent the loss of power over long distances, the voltage is increased. Increasing the voltage reduces the current. This reduces energy loss.

Transformers

Transformers change the voltage of electric current. Current from a power plant goes through a step-up transformer. The current leaves the transformer with a higher voltage. Before entering homes, the current is changed to a lower voltage. A step-down transformer decreases the voltage for your home.

This power plant uses moving water to make electricity. The electricity is sent to homes and businesses through power lines.

The Path of Electrical Energy

① A power plant produces electrical energy.

② A step-up transformer increases the voltage of the electric current.

③ The voltage decreases at a step-down transformer.

④ Another transformer makes the current safe for homes to use.

⑤ Power lines carry electric current back to the power plant.

✓ Quick Check

Look at the picture, and circle the correct answer.

47. What produces the electrical energy?   power plant   transformer

48. What kind of transformer decreases voltage?   step-down   step-up

49. How does current return to the power plant?   transformer   power lines

Energy

Circle the letter of the best answer.

1. The transfer of thermal energy between two objects that are touching is called
   a. convection.
   b. conduction.
   c. evaporation.
   d. heat.

2. The transfer of thermal energy by flowing gases or liquids is called
   a. convection.
   b. conduction.
   c. evaporation.
   d. heat.

3. The height of a wave is its
   a. amplitude. c. wavelength.
   b. echo.   d. volume.

4. A material that lets all the light through so that objects on the other side can be seen clearly is
   a. refraction.
   b. radiation.
   c. transparent.
   d. translucent.

5. A material that completely blocks light from passing through is called
   a. refraction.
   b. opaque.
   c. transparent.
   d. translucent.

6. A material that lets only some light through so that objects on the other side appear blurry is
   a. refraction.
   b. radiation.
   c. transparent.
   d. translucent.

7. The flow of electrical charges through a circuit is
   a. magnetic field.
   b. current electricity.
   c. static electricity.
   d. neutral electricity.

8. The bending of light as it passes from one transparent material into another is
   a. opaque.
   b. translucent.
   c. refraction.
   d. reflection.

Write the answer on the blank. Fill in each blank with one letter.

1. It always moves from warmer objects to cooler objects.

   ___ ___ ___ ___

2. A dolphin can use this to find underwater objects.

   ___ ___ ___ ___

3. You can change this by changing the frequency of a sound wave.

   ___ ___ ___ ___ ___

4. This is made when light falls on a smooth and shiny surface.

   ___ ___ ___ ___ ___ ___ ___ ___ ___ ___

5. A light is on when this is closed.

   ___ ___ ___ ___ ___ ___ ___

6. The Sun's energy reaches Earth this way.

   ___ ___ ___ ___ ___ ___ ___ ___ ___

7. This is the distance between the top of one wave and the next.

   ___ ___ ___ ___ ___ ___ ___ ___ ___ ___

8. This type of wave is caused by vibrations.

   ___ ___ ___ ___ ___ ___ ___ ___ ___

Summarize

_____

_____

_____

_____

Glossary

A

abiotic factor (a·bigh·AH·tic fak·tuhr) any nonliving thing in an environment

absorb (uhb·SORB) to take in or soak up

acceleration (ak·sel·uh·RAY·shun) any change in the speed or direction of a moving object

accommodation (uh·kom·uh·DAY·shun) an organism's response to changes in its ecosystem

acid (AS·id) a substance that turns blue litmus paper red

acid rain (AS·id rayn) harmful rain caused by the burning of fossil fuels

adaptation (a·dap·TAY·shun) a trait or behavior that helps a living thing survive in its environment

air mass (ayr mas) a large body of air that has similar properties throughout

air pressure (ayr PRESH·uhr) the force of air pushing on an area

alternating current (ahl·tuhr·NAYT·ing KUR·uhnt) a current in which electrical charges continuously flow in one direction and then in the opposite direction, AC

alternative energy source (ahl·TUHR·nuh·tiv EN·uhr·jee sors) an energy source other than a fossil fuel

altitude (AL·ti·tewd) the height of a place above sea level

amphibian (am·FIB·ee·uhn) an animal, such as a frog, that spends part of its life in water and the other part on land

amplitude (AM·pli·tewd) the height of a wave

anemometer (an·uh·MOM·i·tuhr) a tool that measures wind speed

apparent motion (uh·PAYR·uhnt MOH·shuhn) the way something seems to move

area (AYR·ee·uh) the number of unit squares that fit inside a surface

astronaut (AS·truh·nawt) a person who travels to space

atmosphere (AT·muhs·feer) the blanket of gases that surrounds Earth

atom (AT·uhm) the smallest part of an element

atomic number (uh·TAHM·ic NUHM·buhr) a number that is related to the mass of an element

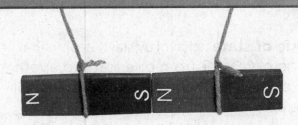

attract (uh·TRAKT) to pull

avalanche (AV·uh·lanch) the sudden downhill movement of a lot of ice and snow

axis (AK·sis) a real or imaginary line that an object spins around

bacteria (bak·TEER·ee·uh) living things that are made up of only one cell, the simplest form of life

balance (BAL·uhns) a tool used to measure mass

balanced forces (BAL·uhnst FOR·suhz) forces that cancel each other out when acting together on a single object

band (band) a thin layer of minerals in a metamorphic rock

barometer (buh·ROM·i·tuhr) a tool that measures air pressure

base (bays) a substance that turns red litmus paper blue

biome (BIGH·ohm) a large ecosystem that has its own kind of climate, soil, and living things

biotic factor (bigh·AH·tic FAK·tuhr) any living thing in an environment

bird (burd) an animal with feathers and wings

boil (boyl) to change from a liquid to a gas by adding thermal energy to the liquid

budding (BUD·ing) a type of reproduction used by some simple invertebrates

buildup (BILD·up) state of having more of one kind of electrical charge than another kind

bulb (bulb) a stem that grows under the ground

camouflage (KAM·uh·flahzh) an adaptation by which an animal can hide by blending in with its surroundings

canopy (KAN·uh·pee) the highest part of a forest ecosystem

canyon (KAN·yun) a steep-sided valley

carnivore (KAHR·nuh·vor) an animal that only eats other animals

cartilage (KAHR·tuh·lij) a tough, flexible material that is similar to bone

cast (kast) a fossil formed or shaped within a mold

cell (sel) the smallest unit of living matter

cell wall (sel wahl) a stiff covering around a plant cell

change of state (chaynj uv stayt) a physical change of matter from one state to another state caused by a change in the energy of the substance

chemical change (KEM·i·kuhl chaynj) a change that results in matter that has different properties than the original matter

chlorophyll (KLAWR·uh·fil) substance found in chloroplasts that helps plants use the Sun's energy to make food

chloroplast (KLAWR·uh·plast) plant cell part that uses sunlight to make food

circuit (SUR·kit) a path along which electric current flows

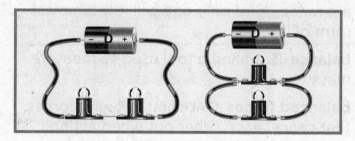

circuit breaker (SUR·kit BRAYK·uhr) a switch that protects circuits from high currents

circulatory system (SUR·kyuh·luh·tor·ee SIS·tuhm) the organ system that moves blood through the body

cirrus cloud (SEER·us klowd) a cloud of ice particles shaped like feathers

class (klas) a smaller classification group within a phylum

classify (KLAS·i·figh) to organize into groups

climate (KLIGH·mit) the average weather pattern of a region over time

climate region (KLIGH·mit REE·juhn) a region of land with certain patterns of temperature, humidity, precipitation, and wind

clone (klohn) an offspring that is an exact copy of its parent

closed circuit (klohsd SIR·kit) an unbroken circuit along which electric current can flow

cloud (klowd) a collection of tiny water droplets or ice crystals that hangs in the air

cold-blooded (kohld BLUD·uhd) an animal whose body temperature depends on its surroundings

cold front (kohld frunt) a cold air mass pushing under a warm air mass

comet (KOM·it) a chunk of ice, rock, and dust that moves around the Sun

community (kuh·MYEW·ni·tee) all the populations in an ecosystem

competition (kom·puh·TISH·uhn) a struggle between organisms for food, water, and other needs

complete metamorphosis (kuhm·PLEET met·uh·MOR·fuh·sis) a life cycle in which each stage is unlike any other stage

compost (KOM·pohst) a mixture of decaying matter that helps plants grow in soil

compound (KOM·pownd) a substance made when two or more elements are joined together and lose their own properties

concave (KON·kayv) curving inward

condensation (kon·den·SAY·shuhn) the process of a gas changing to a liquid

condense (kuhn·DENS) to change from a gas to a liquid

conduction (kuhn·DUK·shuhn) the transfer of thermal energy between materials that are touching

conductor (kuhn·DUK·tuhr) a material through which heat or electricity flows easily

conifer (KON·uh·fur) a plant that makes seeds on cones

conservation (kon·sur·VAY·shuhn) using resources wisely

constellation (kon·stuh·LAY·shuhn) a group of stars that form a pattern in the night sky

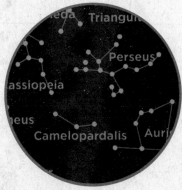

consumer (kuhn•SEW•muhr) an organism that cannot make its own food

continental rise (kon•tin•NEN•tuhl righz) land that connects the ocean floor and continent

continental shelf (kon•tin•NEN•tuhl shelf) the land that connects the shore to the ocean

continental slope (kon•tin•NEN•tuhl slohp) land that slopes down from the continental shelf toward the ocean floor

contour plowing (KON•tor PLOW•ing) a way of plowing fields in curved rows that follow the shape of the land

convection (kuhn•VEK•shuhn) the transfer of thermal energy in gases or liquids

convex (KON•veks) curving outward

cool (kewl) to lose thermal energy

cornea (KORN•ee•uh) a thin, clear tissue on the outside of the eye

crater (KRAY•tuhr) a hollow area or pit in the ground

crust (krust) rock that makes up Earth's outermost layer

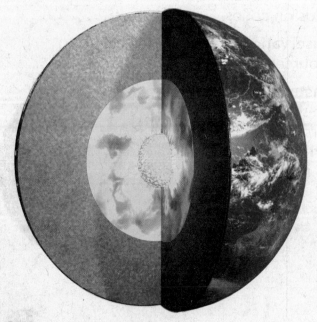

cumulonimbus cloud (KYEW•myu•loh•nim•buhs klowd) a cumulus cloud that has turned dark and thick, causes rain

cumulus cloud (KYEW•myu•luhs klowd) a cloud that is rounded and puffy with a flat bottom

current (KUR•uhnt) a flow of a gas or a liquid

current electricity (KUR•uhnt i•lek•TRIS•uh•tee) the flow of electrical charges through a circuit

cutting (KUT•ing) a part of a plant that has been clipped and can produce a new plant

deciduous forest (di•SIJ•ew•uhs FOR•ist) a forest biome with trees that lose their leaves each year

decomposer (dee•kuhm•POH•zuhr) an organism that breaks down the wastes and remains of other living things

deforestation (dee•for•uh•STAY•shun) the destruction of forest ecosystems

delta (DEL•tuh) a landform that forms where a river meets the ocean

density (DEN•si•tee) the amount of matter in a given space

deposition (de•puh•ZISH•uhn) the dropping off of weathered rock

desert (DEZ·uhrt) a barren biome with very little rainfall

digestive system (digh·JES·tiv SIS·tuhm) the organ system that breaks down food for fuel

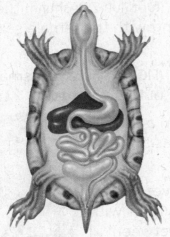

direct current (duh·REKT KUR·uhnt) a current in which electrical charges do not stop or reverse direction, DC

discharge (DIS·chahrj) the sudden movement of electric charges from one object to another object

drainage basin (DRAY·nihj BAY·suhn) the area of land drained by joining rivers

drought (drowt) a period in which less rain falls than usual

dwarf planet (dworf PLAN·it) a small planet often made of rock and ice

earthquake (URTH·kwayk) a sudden shaking of Earth's crust

echo (EK·oh) a sound wave that has bounced off a surface

ecosystem (EE·koh·sys·tuhm) all the living and nonliving things in an environment

efficiency (i·FISH·uhn·see) how well a machine works

effort force (EF·uhrt fors) the amount of force needed to move a load

egg (eg) a female sex cell

electric current (i·LEK·trik KUR·uhnt) a flow of electrical charges

electric motor (i·LEK·trik MOH·tuhr) a device that changes electrical energy into mechanical energy

electromagnet (i·LEK·troh·mag·nit) an electric circuit that makes a magnetic field

electromagnetic spectrum (i·LEK·troh·mag·net·ik SPEK·truhm) the full range of electromagnetic wavelengths

element (EL·uh·muhnt) a substance that is made up of only one type of matter

ellipse (i·LIPS) a flattened circle, or oval

endangered (en·DAYN·juhrd) a living thing that has few of its kind left

endoskeleton (en·doh·SKEL·i·tuhn) an internal support structure in an animal

energy (EN·uhr·jee) the ability to do work, either to make an object move or to change matter

energy pyramid (EN·uhr·jee PEER·uh·mid) the amount of energy at each level of a food web

environment (en·VIGH·ruhn·muhnt) all the living and nonliving things in an area

erosion (i·ROH·zhuhn) the weathering and removal of rock or soil

estuary (ES·chew·er·ee) a place where fresh water flows into salt water

evaporation (i·vap·uh·RAY·shuhn) a liquid changing to a gas

excretory system (EK·skri·tor·ee SIS·tuhm) the organ system that removes wastes from the body

exoskeleton (ek·soh·SKEL·i·tohn) a hard covering that protects the body of some animals

extinct (ek·STINGKT) no longer alive or existing on Earth

family (FAM·i·lee) a smaller classification group within an order

fertilization (fur·ti·luh·ZAY·shuhn) the male sex cell joining with the female sex cell to form a seed

fibrous root (FIGH·bruhs rewt) a small root that spreads out into the soil

filter (FIL·tuhr) a tool that separates things by size

first-class lever (fuhrst klas LEV·uhr) a lever with its fulcrum between the load and the effort force

fixed pulley (fikst PUL·ee) a pulley with its wheel attached to a surface that does not move

flatworm (FLAT·wurm) a simple worm with a flat body

flood (fluhd) a great flow of water over land that is normally dry

food chain (fewd chayn) the path energy takes from one organism to another in the form of food

food web (fewd web) the food chains that are connected in an ecosystem

force (fors) a push or a pull

forecast (FOR·kast) to predict the weather

forest floor (FOR·ist flor) the lowest part of a forest ecosystem

fossil (FOS·uhl) the remains of an organism that lived long ago

fossil fuel (FOS·uhl FYEW·uhl) an energy resource that formed millions of years ago

freeze (freez) to change from a liquid into a solid

frequency (FREE·kwuhn·see) the number of wavelengths that pass a point in a given amount of time

friction (FRIK·shuhn) a force that works against motion

front (frunt) a boundary between air masses with different temperatures

fulcrum (FUL·kruhm) the point on a lever that the bar rotates around

fungus (FUN·gus) a living thing, such as a mushroom, that is similar to a plant but cannot make its own food

fuse (fyewz) a device that helps prevent short circuits by stopping strong electric currents from flowing

G

galaxy (GAL·uhk·see) a very large group of billions of stars

gas (gas) matter that does not have a definite shape or take up a definite amount of space

gas giant (gas JIGH·uhnt) a large planet made mostly of gases, one of the four planets beyond Mars

generator (JEN·uh·ray·tuhr) a device that makes electric current by spinning an electric coil between the poles of a magnet

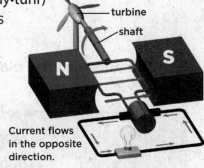
turbine
shaft
N
S
Current flows in the opposite direction.

genus (JEE·nuhs) a smaller classification group within a family

geologic time (jee·oh·LAH·jik TIGHM) a scale used by scientists to measure Earth's history

germination (juhr·muh·NAY·shuhn) when something begins to grow, as when a seed sprouts into a new plant

glacier (GLAY·shuhr) a mass of ice and rock that moves slowly over land

global wind (GLOH·buhl wind) a wind that steadily blows over long distances in a usual pattern

gram (gram) a common unit used to measure mass

grassland (GRAS·land) a biome where the main kind of plant is grass

gravity (GRA·vi·tee) a force of attraction, or pull, between objects

groundwater (GROUND·wa·tuhr) water stored in the cracks and spaces of soil and rock

habitat (HAB·i·tat) the place where a living thing lives

heat (heet) the flow of thermal energy from warmer to cooler objects

herbivore (UR·buh·vor) an animal that eats only producers

heredity (huh·RED·i·tee) the passing of traits from parent to offspring

hibernate (HIGH·buhr·nayt) to rest or sleep through the cold winter

horizon (huh·RIGH·zuhn) a layer of soil

hot spot (hot spot) a place where Earth's crust is over a very hot part of the mantle

humidity (hyew·MID·i·tee) a measure of how much water vapor is in the air

humus (HYEW·muhs) dead plant or animal matter that has broken down

hurricane (HUR·i·kayn) a very large, swirling storm

ice cap (ighs kap) a thick layer of ice on land

igneous rock (IG·nee·uhs rok) rock formed when melted rock cools

imprint (IM·print) a fossil made by a print or an impression

inclined plane (in·KLIGHND playn) a simple machine made of a straight, slanted surface that can multiply an effort force

incomplete metamorphosis (in·cuhm·PLEET met·uh·MOR·fuh·sis) a type of metamorphosis in which a living thing's body form does not change very much

inertia (i·NUR·shah) the tendency of a moving object to keep moving in a straight line

inherited behavior (in·HER·uh·tid bee·HAYV·yur) a set of actions that a living thing is born knowing how to do

inherited trait (in·HER·uh·tid trayt) a trait passed from parent to offspring

inner core (IN·uhr kor) a sphere of solid material at Earth's center

instinct (in·steenkt) an inherited behavior that an animal does without having to be taught

insulator (IN·suh·lay·tuhr) a material that slows or stops the flow of energy, such as heat, electricity, and sound

invertebrate (in·VUR·tuh·brayt) an animal without a backbone

iris (EYE·ris) the colored part of the eye

kilogram (kee·loh·GRAM) a common unit used to measure mass, equal to 1,000 grams

kinetic energy (ki·NET·ik EN·uhr·jee) the energy of motion

kingdom (KING·duhm) the largest group into which a living thing can be classified

landfill (LAND·fil) a place where garbage is put into a lined hole in the ground

landform (LAND·form) a physical feature on Earth's surface

landslide (LAND·slighd) the sudden downhill movement of a lot of rock and soil

larva (LAR·vuh) the wormlike body form of an insect when it has just hatched from its egg

latitude (LAT·i·tewd) how far north or south a location is from the equator

lava (LAH·vuh) melted rock that erupts out of a volcano

law of reflection (law uv ri·FLEK·shuhn) scientific law that states the incoming angle of a light ray is always equal to the outgoing angle

learned behavior (lurnd bee·HAYV·yur) a set of actions that a living thing learns by interacting with its environment or other animals

length (lengkth) the number of units that fit along one edge of an object

lens (lens) a tool that refracts light

lever (LEV·uhr) a simple machine made of a bar and a fulcrum

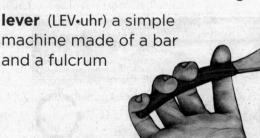

life cycle (lighf SIGH·kuhl) the stages of growth and change that a living thing goes through

life span (lighf span) how long an organism can be expected to live

light (light) a form of energy that lets people see objects

lightning (LIGHT·ning) a discharge of static electricity during a storm

light-year (LIGHT·yeer) the distance light travels in one year, nearly ten trillion kilometers

liquid (LIK·wid) matter that takes the shape of its container and takes up a definite amount of space

litmus paper (LIT·muhs PAY·puhr) a type of paper that can be used to tell if a substance is an acid or a base

load (lohd) the object being lifted or moved by a machine

lunar eclipse (LEW·nuhr i·KLIPS) Earth casting a shadow on the Moon

luster (LUS·tur) the way light bounces off a mineral

machine (muh·SHEEN) anything that helps you do work

magma (MAG·muh) hot, melted rock below Earth's surface

magnet (MAG·nit) an object that has a magnetic field

magnetic attraction (mag·NET·ik uh·TRAK·shuhn) a property that causes a material to be pulled by a magnet

mammal (MAM·uhl) an animal with fur or hair

mantle (MAN·tuhl) the layer of rock below Earth's crust

mass (mas) the amount of matter making up an object

matter (MAT·uhr) anything that has mass and takes up space

melt (melt) to change from a solid to a liquid

metal (MET·uhl) an element that is shiny and conducts heat and electricity

metaloid (MET·uh·loyd) an element that has some but not all of the properties of metals

metamorphic rock (met·uh·MOR·fik rok) rock formed from another kind of rock through heat and pressure

metamorphosis (met·uh·MOR·fuh·sis) a series of changes in a life cycle

meteor (MEE·tee·uhr) a piece of rock, ice, or metal that burns up in Earth's atmosphere, causing a streak of light to appear in the sky

meteorite (MEE·tee·uh·right) a meteor that hits Earth's surface

meteoroid (MEE·tee·uh·royd) a small, rocky body that moves around the Sun

metric system (MET·rik SIS·tuhm) a system of measurement based on units of ten; it is used in most countries and in all scientific work

microorganism (MIGH·kroh·AWR·guh·niz·uhm) a living thing that is too small to be seen with the naked eye

migrate (MYEIGH·grayt) to move from one ecosystem to another as the seasons change

mimicry (MIM·i·kree) the likeness of one living thing to another living thing

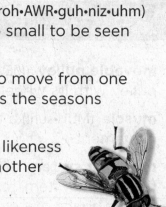

mineral (MIN·uh·ruhl) a natural, nonliving material that makes up rock

mining (MIGHN·ing) digging into the land for useful resources such as minerals, metals, or fuel

mixture (MIKS·chuhr) two or more types of matter that are mixed together but keep their original properties

mold (mohld) an empty space in rock that once held the remains of a living thing

molt (mohlt) to shed the exoskeleton

moraine (muh·RAYN) a hill made from glacial debris

motion (MOH·shuhn) a change in an object's position

mountain (MOWN·tuhn) a tall landform that rises to a peak

movable pulley (MEWV·uh·buhl PUL·ee) a pulley with its wheel attached to the load

muscle (MUH·suhl) a tissue that causes movement

muscular system (MUHS·kyuh·luhr SIS·tuhm) the organ system, made up of muscles, that moves bones

nervous system (NUR·vuhs SIS·tuhm) the master control system of the body

neutral (NEW·truhl) the state a material is in when its electrical charges cancel one another out

newton (NEW·tuhn) a metric unit for weight, measuring an amount of force

nonmetal (non·MET·uhl) an element that has none of the properties of metals

nonrenewable resource (non·ri·NEW·i·buhl REE·sors) a useful material that cannot be replaced easily

nymph (nimf) a young insect

ocean ridge (OH·shuhn ridj) a long mountain range found along the bottom of some oceans

omnivore (OM·nuh·vor) an animal that eats both plants and other animals

opaque (oh·PAYK) completely blocking light from passing through

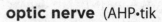

open circuit (OH·pen SIR·kit) a circuit with gaps or breaks

optic nerve (AHP·tik nuhrv) nerve that sends signals from the eyes to the brain

orbit (OR·bit) the path an object takes as it travels around another object

order (OR·duhr) a smaller classification group within a class

organ (OR·guhn) a group of tissues that work together to do a job

organism (OR·guh·niz·uhm) a living thing that carries out five basic life functions on its own

organ system (OR·guhn SIS·tuhm) a group of organs that work together to do a job

ovary (OH·vuh·ree) a part of a plant that stores egg cells

overpopulation (oh·ver·pop·yew·LAY·shun) too many plants or animals living in an area

outer core (OWT·uhr kor) the liquid layer below Earth's mantle

oxygen (OK·suh·juhn) a gas found in air and water that most plants and animals need to live

(P)

parallel circuit (PAYR·uh·lel SUR·kit) a circuit in which an electric current flows through more than one path

periodic table (peer·ee·OD·ik TAY·buhl) a chart that shows the elements classified by properties

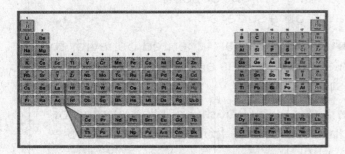

permeability (pur·mee·uh·BIL·i·tee) tells how fast water goes through a porous material

petrified (pe·tri·fighd) turned to stone

phase (fayz) an apparent change in the Moon's shape

photosynthesis (foh·toh·SIN·thuh·suhs) a process in plants that uses energy from sunlight to make food from water and carbon dioxide

phylum (FIGH·luhm) a smaller classification group within a kingdom

physical change (FIZ·i·kuhl chaynj) a change that begins and ends with the same type of matter

physical weathering (FIZ·i·kuhl WETH·uhr·ing) a slow process that breaks rocks into smaller pieces without changing the minerals that the rocks are made of

pistil (PIS·tuhl) a female plant part that makes eggs

pitch (pitch) the highness or lowness of a sound

plain (playn) a large area of land with no hills or mountains

planet (PLAN·it) a round object in space that travels around the Sun

pole (pohl) one of two ends of a magnet, where a magnet's pull is strongest

pollen (PO·luhn) a powder some plants use when reproducing

pollination (pol·uh·NAY·shuhn) the process of moving pollen

pollution (puh·LEW·shuhn) harmful or unwanted material that has been added to the environment

population (pop·yuh·LAY·shuhn) all the members of a species that live in an ecosystem

pore space (por spays) the space between particles of soil

position (puh·ZISH·uhn) the location of an object

potential energy (puh·TEN·shuhl EN·uhr·jee) stored energy

prairie (PRAYR·ee) a grassland with a mild climate

precipitation (pree·sip·uh·TAY·shuhn) water that falls from clouds down to Earth

predator (PRED·uh·tuhr) a living thing that hunts other organisms for food

prey (pray) a living thing that is eaten by other organisms

prism (PRIZ·uhm) an object that separates white light into bands of colored light

probe (prohb) an unmanned spacecraft that leaves Earth's orbit

producer (pruh·DEW·suhr) an organism that makes its own food using energy from sunlight

property (PROP·uhr·tee) a characteristic of matter you can observe, such as color, shape, and size

pulley (PUL·ee) a simple machine made of a grooved wheel with a rope in the groove

pupa (PYEW·puh) the insect stage of growth where the larva develops into an adult

pupil (PYEW·puhl) the hole in the center of the eye

radiation (ray·dee·AY·shuhn) the transfer of thermal energy through space

rain gauge (rayn gayj) a tube used to measure how much rain falls

ray (ray) a straight line that extends away from a point

recycle (ree·SIGH·kuhl) to make a new product from old materials

reduce (ri·DEWS) to use less of something

reflection (ri·FLEK·shuhn) the bouncing of light or sound waves off a surface

reflex (REE·fleks) an inherited behavior that an animal does as an automatic reaction to a stimulus

refraction (ri·FRAK·shuhn) the bending of light as it passes from one material to another

regeneration (ree·jen·uh·RAY·shuhn) the process of a complete animal growing from just one part of the original animal

relative age (REL·uh·tiv ayj) the age of one thing as compared to another

renewable resource (ri·NEW·i·buhl REE·sors) a useful material that is replaced quickly in nature

repel (ri·PEL) to push away

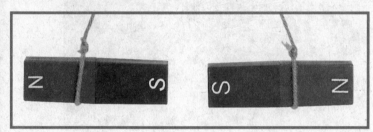

reproduce (ree·pruh·DEWS) to make offspring

reproduction (ree·pruh·DUK·shuhn) the making of new living things

reptile (REP·tyel) an animal, such as a lizard, that has dry skin covered in scales or plates

reservoir (REZ·uhr·vwahr) a storage area for holding fresh water

resistance (ri·ZIS·tuhns) the ability to slow electric current

respiratory system (RES·pruh·tor·ee SIS·tuhm) the organ system that brings oxygen to body cells and removes carbon dioxide

retina (RET·uh·nuh) the part of the eye that changes light images into nerve signals

reuse (ree·YEWZ) to use something over again

revolution (REV·uh·LEW·shuhn) one complete trip around an object in a circular or nearly circular path

rock cycle (rok SIGH·kuhl) the process in which rocks change from one type to another

rocky planet (ROK·ee PLAN·it) a planet made mostly of rock, one of the four closest planets to the Sun

root (rewt) the part of a plant that takes up water and minerals from the ground

rotation (roh·TAY·shuhn) the spinning of an object around its axis

roundworm (ROWND·wurm) a worm with a thin body and pointed ends

runoff (RUN·awf) water that flows over the surface of the land

rust (rust) a solid brown compound that forms when iron combines chemically with oxygen

S

sand dune (sand dewn) a hill made of sand

satellite (SAT·uh·light) a natural or artificial object that circles another object in space

savanna (suh·VAN·uh) a grassland with shrubs and few trees

screw (skrew) a simple machine made of an inclined plane wrapped around a bar

second-class lever (SEK·uhnd klas LEV·uhr) a lever with its fulcrum at the end and its load in the middle

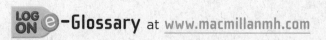

sediment (SED·uh·muhnt) bits of soil or rock that may be eroded and deposited

sedimentary rock (sed·uh·MUHN·tuh·ree rok) rock formed from sediments that are stuck together

seed (seed) a plant that is not fully formed

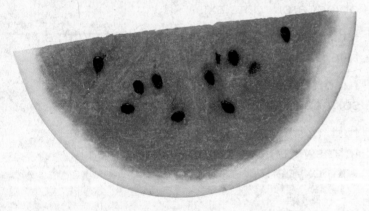

segmented worm (SEG·ment·uhd wurm) a worm with a body divided into parts, or segments

seismic wave (SIGHZ·mik wayv) a vibration caused by an earthquake

seismograph (SIGHZ·muh·graf) a tool that finds and records earthquakes

series circuit (SEER·eez SUR·kit) a circuit in which an electric current flows in the same direction along a single path

settling (SET·ling) a way of separating a mixture by letting denser material sink to the bottom

short circuit (short SUR·kit) a circuit with little resistance to slow an electric current

simple machine (SIM·puhl muh·SHEEN) something that has only a few parts and makes it easier to do work

skeletal system (SKEL·i·tuhl SIS·tuhm) the organ system, made up of bones, that supports the body

soil profile (soyl PROH·fighl) a view of the different layers in a soil sample, from the surface down to the bedrock

solar eclipse (SOH·luhr i·KLIPS) the Moon casting a shadow on Earth

solar system (SOH·luhr SIS·tuhm) the Sun and all the objects that travel around it

solid (SOL·id) matter with a definite shape that takes up a definite amount of space

solution (suh·LEW·shuhn) a mixture in which one or more types of matter are mixed evenly in another kind of matter

sound wave (sownd wayv) a wave that travels along or through matter, produced by a vibration

species (SPEE·sheez) a group of similar living things that reproduce more of their own kind

speed (speed) the distance traveled in an amount of time

sperm (spurm) a male sex cell

spore (spor) a cell in a seedless plant that can grow into a new plant

stamen (STAY·muhn) a male plant part

standard unit (STAN·duhrd YEW·nit) a measuring unit commonly used in the United States, such as inches, pounds, and ounces

star (stahr) a sphere of hot gases that gives off light and heat

state of matter (stayt uv MAT·uhr) the form that matter is in; solid, liquid, or gas

static electricity (STAT·ik i·lek·TRIS·i·tee) the buildup of charged particles

stationary front (STAY·shuh·NER·ee frunt) a boundary between air masses that are not moving

stem (stem) the part of the plant that holds the plant up and moves water and nutrients to and from the roots and leaves

stimulus (STIM·yuh·luhs) *sing.*, **stimuli** (STIM·yuh·ligh) *pl. n.*, something in an environment that causes a living thing to respond

stoma (STOH·muh) *sing.*, **stomata** (stoh·MAH·tuh) *pl. n.*, tiny openings on a plant leaf that open or close to let in air and give off water vapor

stratus cloud (STRAT·us klowd) a cloud shaped like layers of blankets

streak (streek) the mark left behind when you scratch a mineral against a white tile

subsoil (SUB·soyl) a hard layer of clay and minerals beneath topsoil

survive (sur·VIGHV) to continue to live

switch (swich) a device that turns current electricity on and off

symmetry (SIM·uh·tree) the matching up of body parts around a midpoint or line

(T)

taiga (TIGH·guh) a cool forest biome in the far north

taproot (TAP•rewt) the main root of plants such as carrots; a thick, pointed root that goes straight down into the ground

telescope (TEL•uh•skohp) a tool that makes objects look closer and larger

temperature (TEM•puhr•uh•chuhr) how hot or cold something is

terminus (TUR•min•us) the end of a glacier

thermal energy (THUR•muhl EN•uhr•jee) the energy of moving particles of matter

thermometer (thuhr•MOM•i•tuhr) a tool that measures temperature

third-class lever (thurd klas LEV•uhr) a lever with its fulcrum at the end and the effort force is in the middle

tide (tighd) the regular rise and fall of the ocean's surface

tissue (TISH•yew) a group of cells that do the same job

topsoil (TOP•soyl) the dark, top layer of soil, rich in humus and minerals, in which many organisms live and grow

tornado (tor•NAY•doh) a column of spinning wind that moves across the ground in a narrow path

trait (trayt) a feature of a living thing

transfer (trans•FUHR) to pass from one object to another

transform (trans•FORM) to change from one form to another

transformer (trans•FORM•uhr) a device that changes the voltage of electric current

translucent (trans•LEW•suhnt) letting only some light through so that objects on the other side appear blurry

transparent (trans•PAYR•uhnt) letting all the light through so that objects on the other side can be seen clearly

tropical rain forest (TROP•i•kuhl rayn FOR•ist) a biome that is hot and humid with a lot of rainfall

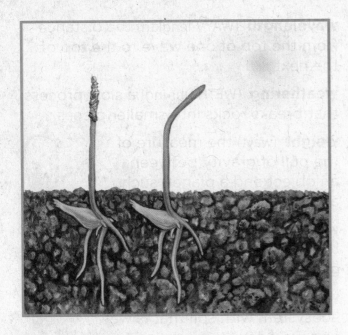

tropism (TROH·pi·zum) a plant's response to a stimulus

tropisphere (TROHP·uh·sfeer) the layer of the atmosphere that is closest to Earth's surface

tsunami (SEW·NAH·mee) a large ocean wave that is often caused by an earthquake under the ocean

tuber (TEW·bur) a storage part of a plant

tundra (TUN·druh) a cold, dry biome with frozen ground and no trees

unbalanced forces (un·BAL·uhnst FOR·suhz) forces that do not cancel each other out when acting together on a single object

understory (UN·dur·stor·ee) the middle part of a forest ecosystem, made up of young trees and shrubs

velocity (vuh·LOS·i·tee) the speed and direction of a moving object

vertebrate (VUR·tuh·brayt) an animal with a backbone

vibration (vigh·BRAY·shuhn) a back-and-forth motion

virus (VIGH·rus) a particle that uses a living cell to reproduce; causes disease

visible spectrum (VIZ·uh·buhl SPEK·truhm) the part of the electromagnetic spectrum that human eyes can see

volcano (VOL·kay·noh) a mountain of once molten rock formed around an opening in Earth's crust

volt (vohlt) a unit of measurement for the strength of a power source

volume (VOL·yewm) 1. the amount of space an object takes up; 2. the loudness or softness of a sound

warm-blooded (wawrm BLUD·uhd) an animal whose body temperature does not change much

warm front (wawrm frunt) a warm air mass pushing into a cold air mass

water cycle (WA·tuhr SIGH·kuhl) the movement of water between Earth's surface and the air

watershed (WA·tuhr·shed) an area of land where water flows downhill to a common stream, lake, or river

water treatment plant (WA·turh TREET·muhnt plant) a facility that people use to make water clean

water vapor (WA·turh VAY·puhr) water in the gas state

wavelength (WAYV·lengkth) the distance from the top of one wave to the top of the next

weathering (WETH·uhr·ing) a slow process that breaks rocks into smaller pieces

weight (wayt) the measure of the pull of gravity between an object and a planet, such as Earth

well (wel) a hole dug below the ground in order to reach groundwater

wetland (WET·land) a land ecosystem with soil that is wet or covered with water during part or all of the year

wheel and axle (wheel and AK·suhl) a simple machine made of a wheel and a smaller axle

width (width) the number of units that fit across an object

wind vane (wind vayn) a tool that points in the direction that wind is blowing

work (wurk) the use of force to move an object a certain distance

Credits